CROP DISEASES
IDENTIFICATION AND MANAGEMENT

A COLOUR HANDBOOK

CROP DISEASES
IDENTIFICATION AND MANAGEMENT

A COLOUR HANDBOOK

L.V. Gangawane
V.C. Khilare

2024
Daya Publishing House®
A Division of
Astral International Pvt. Ltd.
New Delhi – 110 002

First Published, 2008

Reprinted, 2024

ISBN: 978-81-7035-539-7 (HB)

Published by : **Daya Publishing House®**
A Division of
Astral International Pvt. Ltd.
– ISO 9001:2015 Certified Company –
4736/23, Ansari Road, Darya Ganj,
New Delhi-110 002
Ph. 011-43549197, 23278134
E-mail: info@astralint.com
Website: www.astralint.com

Printed at : **Replika Press Pvt. Ltd.**

डा. चारू दत्त मायी
अध्यक्ष
Dr. C.D. Mayee
CHAIRMAN

कृषि वैज्ञानिक चयन मंडल
भारतीय कृषि अनुसंधान परिषद
कृषि अनुसंधान भवन-1, पूसा,
नई दिल्ली - 110 012

ICAR

AGRICULTURAL SCIENTIST RECRUITMENT BOARD
INDIAN COUNCIL OF AGRICULTURAL RESEARCH
KRISHI ANUSANDHAN BHAVAN-1, PUSA,
NEW DELHI – 110 012
Telephone: (O) 25843295, 25846540
Fax: 91-11-25846540 Telegram: AGRECBOARD
E-mail: cdmayee@icar.org.in

Forword

A handbook has several cannotations. To some it means journal, to others it means log, chronicle, memoirs, record, writing and so on. When Dr. L.V. Gangawane and V.C. Khilare put this compilation of "*Crop Diseases: Identification and Management– A Colour Handbook*", it basically meant a guide and a brief record of the keys of identification and management of crop diseases. Why was this a necessity when volumes of literature are available on the subject? When the concept emerged of preparing the handbook on this subject two things were clearly behind the project. The first one was farmers' need. They grow variety of crops in a cropping system mode and hence booklets and keys specially designed for crops were of limited use. They array of crops included in the text serve the desired purpose. The second necessity was quick guide to broad symptoms and management steps. Therefore, the handbook is designed for farmers for self help in identification of disease malady of various crops with the help of pictures and follow immediate strategies of management. Once the farmer adopts these practices, he or she never looses time that is critical is management. Thus handbook in a teacher and will certainly be an asset in the toolbox of an Indian farmer.

The author deserves congratulations for doing an excellent job through involving large number of crop pathologists who are specialized in their areas and bringing very useful information for education of the growers.

(C.D. Mayee)
Chairman,
Agricultural Scientist Recruitment Board

Preface

It is fortunate that we are inspired by the farmers of our country to provide such a simple diary. Our experience is that farmers are unable to identify the disease of crop plants in the field and apply the correct method of disease management although now-a-days tremendous improvement in identification and management have been investigated by the plant pathologist. Because of the production of food grains has increased from 50 million tones to more than 210 million tones and projected demand in 2020 shall be more than 300 million tons. It is said that there are nearly 5000 diseases of crop plants in India. Some times there are epidemics due to failure of resistance.

In this diary, the efforts are made to identify diseases visually by the farmers and manage the disease timely. Now-a-days Hon'ble Agricultural Minister, Govt. of India is announcing to educate the farmers regarding the integrated management of crops. The losses of crop due to diseases and pests are most important. We are highly indebted to Hon'ble Shri Sharad Pawar, Agricultural Minister, Govt. of India for announcing this type of programme and we have planned to translate this diary in different Indian languages through different scientists all over India. We are highly grateful to Dr. Y. L. Nene, Chairman, Asian Agri History Foundation, Secunderabad and Dr. C. D. Mayee, Chairman, ASRB, New Delhi who has constantly encouraged us for this task and also provide information on some tomato diseases. The diary would not have completed without the help of pathologist from various ICAR crop centers. We are grateful to them. We also thank Daya Publishing House, New Delhi for bringing this diary for farmers within stipulated time.

L.V. Gangawane

V.C. Khilare

Contents

Bajra

Banana

Betelvine

Black Pepper

Cardamom

Chickpea

Chilli

Citrus

Cocoa

Coconut

Groundnut

Guava

Mango

Pomegranate

Potato

Rice

Soybean

Sugarcane

Sunflower

Tomato

Wheat

Apple

Apple Scab

Vijay Singh Thakur

**Regional Horticultural Research Station,
YSPUHF, Mashobra, Shimla (HP)**

Symptoms

Apple is grown throughout the temperate region of the world. In Indian Himalayas, the cultivation is extended from Kashmir to Northeastern hills, occupying a prime position among various temperate fruits and accounting for 55 per cent of the total area and 75 per cent of the total production. Apple is grown as perennial monocultures, the disease problems are entirely different and complex in nature. Such diseased situations have led to repeated and excessive use of chemical pesticides, which has resulted in the development of resistance in the pathogens and environmental pollution. Over 96 diseases have been reported to affect the apple crop resulting in reduction in yield, devaluation in grade, foliage loss and operational expenditure. The magnitude of control through fungicides can be gauged from the fact that in one season 8-12 sprays are given by the farmers on apple entailing an expenditure of Rs 32000-40000/ha.

Briefly, this note is prepared on three major foliar diseases, which caused heavy losses to apple plant and productivity during last two decades onwards.

Pathogen

Venturia inaequalis

Management

Integrated and monitored spray schedule based on plant pathology and weather conditions has been devised for the control of apple scab disease during spring for primary infection/symptoms and summer for secondary infection/symptoms on leaves and fruits. This annually revised spray schedule includes preventive, curative and eradicative types of fungicides to combat the disease every year. Overwintering sprays of 5 per cent urea a foliar nitrogenous fertilizer at leaves shedding stage and followed by a spray of Bordeaux mixture (1 per cent) are most effective IDM strategy used to eradicate the most overwintering pathogens/diseases on apples. The control of apple scab disease is a success story to bring the infection to eradication stages in last five years. However, the scab disease studies reveal its inconsistent nature from epidemic to sporadic, hence the technology claim is made and documented to the apple orchardists as "scab on apples is not a disease which can be eradicated as with the advent of wet weather, it may reappear in the orchards to the epidemic level even after absent for last many years but can be managed. Thus orchardists are required to keep vigil of weather conditions and carry on necessary control measures to keep the disease under check.

Apple Blotch

Vijay Singh Thakur

Regional Horticultural Research Station,
YSPUHF, Mashobra, Shimla (HP)

Symptoms

Pathogen

The disease is caused by *Marssonina coronaria* and *Alternaria mali* a new fungal disease is predominant and widespread on apple leaves and fruits. The damage was up to 64 per cent on leaves as premature defoliation and 12-15 per cent on matured fruits. The disease first attacked plants growing in shady and high moisture conditions in the orchards where air temperature, light intensity and soil moisture are the predisposing factors to trigger the disease outbreak, however, severity index increases with the maturity of the leaves. It is most severe when a cool, rainy period of 4-5 days proceeds several hot sunny days in the orchard. It is also obvious that the inoculum of the disease get released from the pink bud stage but it causes the infection only at a particular plant physiological stage *i.e.* when leaves start producing ethylene, the ripening/ageing hormone. The pathogen therewith the hormonal changes, induces yellowing and abscission of the leaves and defoliation occurs.

Management

Mixture of dithiocarbamates (0.3 per cent) and carbendazim (0.01 per cent) at 15-20 days interval was developed to manage the disease. The fungicides sprays must start from the first week of June in whatever the weather conditions are prevailing in orchards. This strategy is effective to manage the disease at few sprays to the end of September month every year.

Apple Powdery Mildew

Vijay Singh Thakur

Regional Horticultural Research Station, YSPUHF, Mashobra, Shimla (HP)

Symptoms

Pathogen

Powdery mildew (*Podosphera leucotricha)* is a third major foliar disease of apple and prevalent in all apples growing states. Severe infection under high humidity conditions leads to deformation of mildew-affected shoots with subsequent curtailment of its growth and a reduction in the number of potential fruiting buds. Maximum damage is at the flowering time. The conidia infest the blossom parts and damages the flowering and fruit set. The most congenial infection conditions are at relative humidity of more than 90 per cent and temperature ranging between 19 to 22°C.

Management

The disease infection conditions can be monitored and lesser fungicidal sprays can be recommended. Specific fungicides like Penconazole (0.05 per cent), Hexaconazole (0.05 per cent), Difenoconazole (0.015 per cent) and Carbendazim (0.01 per cent) at short interval though can eradicate powdery mildew to larger extent but uneconomical in terms of net losses. In fact, fungicides on powdery mildew can only kill the germinated conidia and not the remaining conidia till their germination. Integrated sprays of fungicides at pink bud and petal fall stages are recommended on apples to manage symptoms of the disease every year.

Arecanut

Fruit Rot of Arecanut

M. Gunasekaran and R. Chandramohanan

Central Plantation Crops Research Institute,
Kasaragod – 671 124, Kerala

Symptoms

1. Water soaked lesions appear near the perianth.
2. Later these lesions spread over other parts giving the nut a dark green colour (Figure 1).
3. Infection causes nut shedding.

Figure 1

Pathogen

Phytophthora arecae/P. meadii.

Management

1. Spraying with 1 per cent Bordeaux mixture immediately after the first monsoon showers and the second round of spray after 40-45 days.
2. Spraying operations are to be undertaken on clear sunny days. Diseased nuts and bunches are to be destroyed to reduce the inoculum.
3. Polythene covering (125-200 gauge 24 x 30 inches) can as well substitute the traditional method of spraying Bordeaux mixture for effective management of kole roga in arecanut.

Bud Rot of Arecanut

M. Gunasekaran and R. Chandramohanan

**Central Plantation Crops Research Institute,
Kasaragod – 671 124, Kerala**

Symptoms

1. The disease was recorded in a severe form in the heavy rainfall tracts of Karnataka.
2. Though the disease generally occurs in monsoon season, the fresh infection during November onwards becomes severe during succeeding months.
3. The crops loss due to this disease is 1 per cent or more.
4. Tender leaf becomes yellow, which later changes to brown (Figure 1).
5. Spindle can be drawn out of the crown with a slight pull in advanced stages.
6. Infection spreads to outer whorl of leaves leading to their yellowing, drooping and shedding (Figure 2).
7. Diseased palm emits an odd smell.

Figure 1

Figure 2

Pathogen

Pytopthora arecae/P. meadii

Management

1. Cut and remove the rotten tissues and burn.
2. Apply Bordeaux paste (10 per cent) cover it with polythene sheet.
3. The disease can be prevented by prophylactic spray with Bordeaux mixture (1 per cent) during pre and post monsoon period.

Ganoderma or Anabe of Arecanut

M. Gunasekaran and R. Chandramohanan

Central Plantation Crops Research Institute, Kasaragod – 671 124, Kerala

Symptoms

1. Yellowing of outer whorl leaves, which gradually extend to inner whorls.
2. Leaves show wilting symptoms and one by one droops down.
3. Small brown irregular patches appear on the basal portion of the stem and dark brown liquid oozes, in the advanced stage of infection (Figure 1).
4. Infected roots are discoloured, brittle, dry and give nasty smell.
5. Development of inflorescence is arrested. The nuts already formed are shed.
6. Leaves shed one by one and the weakened crown topples leaving a bare trunk. The fruit body of the fungus (anabe) appears at the base of the stem (Figure 2).
7. The disease is fatal and affects roots and crown. The disease is a major problem in the Malnad or Maidan areas of Karnataka and also parts of Tamil Nadu, Kerala and Assam.

Pathogen

Ganoderma lucidum/G. applanatum.

Figure 1

Figure 2

Management

1. Diseased palms in the early stages may be root fed with 125 ml of 1.5 per cent Calixin at quarterly intervals.
2. The basin of the affected palms may be drenched with 1.5 per cent Calixin at quarterly intervals.
3. Affected palms may be given 2 kg of neem cake/palm.
4. The stem of all diseased palms should be removed along with roots and burnt.
5. Palms in the different stages of the disease should be isolated from healthy palms by digging trenches of 30 cm width and 60 cm depth.
6. The health of the palms may be improved by applying normal dose of NPK fertilizers.
7. Repeated ploughing, water stagnation and closer spacing may be avoided.

Yellow Leaf of Arecanut

M. Gunasekaran and R. Chandramohanan

Central Plantation Crops Research Institute,
Kasaragod – 671 124, Kerala

Symptoms

1. Characteristic yellowing of leaves starts from the tip and progress inwards (Figure 1). Vein banding is seen (Figure 2).
2. Yield of the palm decreases; kernel shows discoloration and turn blackish.
3. Root tips become dark and rot.
4. The crown size reduces.

Figure 1

Figure 2

Pathogen

Phytoplasma (Figure 3). Electronmicrograph showing pleomorphic bodies of phytoplasma in the phloem cells.

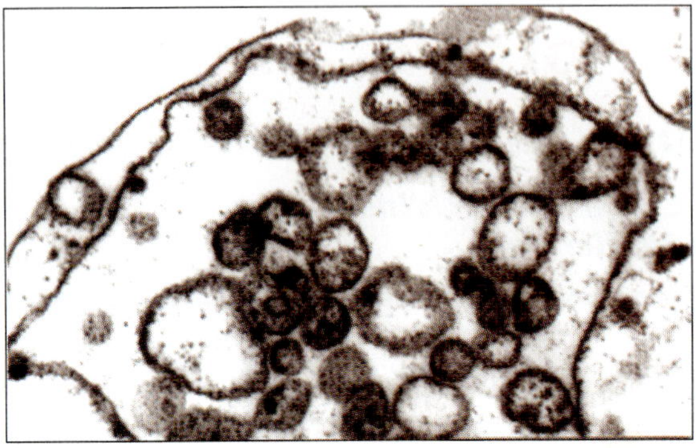

Figure 3

Management

In Heavily Affected Area

The following management practices have been recommended to improve the yield of YLD affected palms:

1.. Regular and balanced fertilizer application. Apply 200 g of urea, 200 g mussorie phos and 200 g of muriate of potash per palm per year in two split doses.
2.. In addition to the above, apply 800 g of mussorie phos per palm.
3. Apply organic manures @ 12 kg each of compost and green leaves per palm per year.
4.. Provide irrigation during summer months.
5. Avoid water stagnation by providing drainage facilities.
6. Grow cover crops in the garden.
7. Adopt need based plant protection measures against other pest and diseases.
8. Loss can be reduced to the minimum if the palms could be attended immediately on appearance of symptoms.

When only a few palms are affected in a garden, remove them to delay further spread of the disease within the garden.

In Mildly Affected Areas

The strategy for mildly affected areas is to contain the disease by removing all the diseased palms, continue monitoring the palms and remove as soon as symptoms appear.

Inflorescence Die-back and Button Shedding of Arecanut

M. Gunasekaran and R. Chandramohanan

Central Plantation Crops Research Institute, Kasaragod – 671 124, Kerala

Symptoms

1. Yellowing and drying of the rachis from the tip downwards (Figure 1).
2. Shedding of the female flowers and buttons and complete drying of the inflorescence.

Figure 1

3. The disease affects the female flowers and developing buttons, leading to the reduction in the yield.
4. It is prevalent in the Kerala and Karnataka and is severe during summer months from February to May.

Pathogen

Colletotrichum gloeosporioides.

Management

1. Spraying with Dithane Z 78 @ 4g/litre of water or combination of copper sulphate and aureofungin sol both at 50 ppm. First spraying at the time of opening of female flowers and the second 20-25 days thereafter.
2. Remove and destroy all diseased dried bunches as a phytosanitory measure to reduce the inoculum.

Bajra

Downy Mildew of Bajra

V.V. Datar and Sunita J. Magar

Department of Plant Pathology, Marathwada Agricultural University, Parbhani – 431 402

Symptoms

Leaves of infected plants become light green to yellow in colour On lower side of infected leaves there is whitish growth of the fungus. The infected plants are dwarf as compared to healthy ones. Produce excessive fillers leaves become shredded and brown. In green ear stage in place of grains hairy green structures are produced in large number and no grains are produced. Partial infection of ear head is observed in which some grains are produced and rest of the earhead is converted in to a green ear.

Pathogens

Sclerospora graminicola.

Management

Inter cropping of Pearl millet and Pigeon Pea in 4:2 ratio will hinder the secondary spread of the pathogen. Deep plouging reduces population of pathogen in soil. Use of seed from disease free crop.

1. Resistant Varieties

Pearl millet population like ICTP-8203, AIMP-92901 (Samrudhi), PPC-6 (Sampada) and hybrids like Shradha, Saburi and Pratibha (AHB-1666) are resistant to downy mildew.

2. Alteration of Crop Practices

Application of 25 kg N/ha 25 day after sowing boosts crop growth and the crop escapes from the attack of downy mildew.

3. Biological Control

Seed treatment with *Trichoderma viride or T.harzianum* or *Pseudomonas fluorescence* preparation @ 10 g/kg of seed can reduce the incidence of downy mildew.

4. Chemical Control

Seed treatment with metalaxyl (Apron 35 S.D.) @ 6 gm/kg of seed can effectively control the downy mildew.

5. Integrated Management

1. Use of resistant variety or hybrid
2. Seed treatment with Metalaxyl (Apron 35SD) @ 6 gm/kg of seed.
3. Rouging of infected plants.
4. Crop rotation.

Rust of Bajra

V.V. Datar and Sunita J. Magar

Department of Plant Pathology, Marathwada Agricultural University, Parbhani – 431 402

Symptoms

The minute, round uredosori occur in groups on both surfaces of leaves, particularly towards the distal end of the blade, on the leaf sheath and stem. The telia, which appear later in the season are black, elliptical and sub epidermal and are also distributed over the leaf blade, leaf sheath and stem. The plant appears unhealthy and slightly stunted, compared with healthy plant.

Pathogen

Puccinia pennisetti.

Management

This disease not occurs in severe condition. More recently preventive spray with 100 ppm of cupramar and dithane S31 found effective.

Leaf Blast of Bajra

V.V. Datar and Sunita J. Magar

Department of Plant Pathology, Marathwada Agricultural University, Parbhani – 431 402

Symptoms

Symptoms on the leaves originate as small specks, which subsequently enlarge into spindle shaped specks length 0.5cm to several cms. The center of a well-developed spot is whitesh grey with a broken margin. Symptoms also appeared on stem. In severe cases black leaves show blast symptoms.

Pathogen

Pyricularia setariae.

Management

Spray mancozeb 75 per cent

Ergot of Bajra

V.V. Datar and Sunita J. Magar

Department of Plant Pathology, Marathwada Agricultural University, Parbhani – 431 402

Symptoms

Small droplets of a light, honey coloured dew-like substance exude from infected spikelets. A few to many such spikelets may be found in-group, which darken with age, and small grayish or dark brown scelerotia are formed.

Pathogen

Claviceps microcephala.

Management

Primary infection through sclerotial seed. Infected seed should be dipped in 30 per cent salt solution separates out the sclerotia and sclerotial fragments from the grains.

1. Alteration of Crop Practices

Disease occurs when flowering comes at rainy season at that time humidity is high so adjusting the date of sowing as flowering not concide with rains. Destruction of alternate host *Cenchurus cilaris* and *C. seligerus* and wild *Permisetun* sp.

2. Biological Control

Spray of *Allium Sativum* bulb extract is reported to reduce the disease.

3. Chemical Control

Spray with ziram, a mixture of copper oxychloride and zineb (0.2) applied 2–3 times at weekly interval.

Smut of Bajra

V.V. Datar and Sunita J. Magar

Department of Plant Pathology, Marathwada Agricultural University, Parbhani – 431 402

Symptoms

A few grains, sporadically distributed on the ear, may be replaced by oval to top shaped sori, which are generally two or three times the size of the normal grain. Such seeds are bright green to dirty black in colour and often rapture to expose a black spore mass.

Pathogen

Tolyposporium penicillariae

Management

1. Deep ploughing in hot weather reduces population survive in soil.
2. Field sanitation.
3. Crop rotation.
4. Removal of smutted ears.
5. Use of clean seed.

1. Alteration of Crop Practices

Intercropping of Mungbean with Pearl millet.

2. Resistant Varieties

Certain genotypes SSC PS – 252-54 and ICI 7517-91 show high level of resistance.

3. Integrated Management

1. Use of resistant varieties/hybrid.
2. Field sanitation.

Banana

Panama Wilt of Banana

V.C. Khilare

Vasantrao Naik Mahavidyalaya, Aurangabad – 431 003

Symptoms

1. This disease is commonly found in poorly drained soils.
2. Initial symptoms are yellowing of lower leaves, including leaf blades and petioles.
3. The leaves hang around the pseudostem and wither. In the pseudostem of the diseased plant, yellowish to reddish streaks are noted with intensification of colour towards the rhizome.
4. Wilt is severe in poor soil with continuous cropping of banana.
5. Warm soil temperature, poor drainage, light soils and high soil moisture are congenial for the spread of the disease.

Pathogen

Fusarium oxysporum f.sp. *cubens.*

Management

1. Severely affected plants should be uprooted and burnt.
2. Highly infected soil should not be replanted with banana at least for 3-4 years.
3. Use of disease-free planting material and resistant cultivar are recommended.

4. Growing of paddy followed by banana for 3-5 years once or twice, use of quick lime near the base of the plant and soaking with water and avoiding sunflower or sugarcane in crop rotation helps to reduce the disease incidence.
5. Dipping of suckers in carbendazim (10g/10 litres of water) followed by bimonthly drenching starting from 6 months after planting is also recommended.
6. Application of bioagents, such as, *Trichoderma viride* or *Pseudomonas fluorescence* in the soil is effective.

Sigatoka Disease of Banana (Yellow and Black)

V.C. Khilare

Vasantrao Naik Mahavidyalaya, Aurangabad – 431 003

Symptoms

1. Yellow sigatoka is one of the serious diseases affecting the banana crop.
2. Initial symptoms appear in the form of light yellowish spots on the leaves.
3. A small number of these enlarge, become oval; the colour also changes to dark brown.
4. Still later, the centre of the spot dies, turning light grey surrounded by a brown ring.

5. In severe cases, numerous spots coalesce, killing large parts of the leaf.
6. Rainfall, dew and temperature determine the spread of the disease. Conditions favouring mass infection are most common during the rainy season with temperature above 21°C.

Pathogen

Yellow sigatoka: *Mycosphaerella musicola*

Black sigatoka: *Mycosphaerella fijiensis*

Management

1. Cultural practices such as improved drainage, control of weeds, removal of diseases suckers and adopting correct spacing is recommended.
2. Dithane M-45 WP (in oil-water emulsion) and dithane M-45 (in water only) controlled this disease.
3. Foliar spray of copper oxychloride (3 g/litre of water) or thiophanate methyl 1 g/litre of water) controls the disease effectively.

Betelvine

Foot Rot/Leaf Rot of Betelvine

V.V. Datar and Sunita J. Magar

Department of Plant Pathology,
Marathwada Agricultural University, Parbhani – 431 402

Symptoms

The first symptom of the disease is darkening of the basal stem. Soon the vine wilts, accompanied by yellowing of leaves and dropping of the vines, starting from the tip downwards. Closer examination of the plant reveals the spread of the infection from the soil level upwards, reaching the second or third internode. The roots of the affected plants are discoloured, rotted, starting from the smaller roots and spreading to the stem.

In case of leaf rot, the first symptoms are the appearance of water soaked lesions, often starting from the edge and at times towards the centre without a well-defined margin. The infected spots enlarge rapidly to cover a port or whole of the blade, which starts retting. The diseased leaves turn brown to dark brown and later dirty black. As the disease advances the infection spreads to the petiole and stem.

Pathogen

Phytopthora parasitica

Management

Spraying with 0.5 per cent Bordeaux mixture, Peronox (0.35 per cent), Faytolan (6 to 10 gallons), Dithane Z-78 (0.2 per cent), three to four times at 8-10 days interval. Drenching soil with 1 per cent Bordeaux mixture. Drenching with cashewnut compound or adding lime, sulphur or other chemicals and mixing thoroughly, sterilizes soil.

Wilt of Betelvine

V.V. Datar and Sunita J. Magar

Department of Plant Pathology,
Marathwada Agricultural University, Parbhani – 431 402

Symptoms

Initial symptoms start with yellowing and drooping of leaves. Gradual wilting of vines is observed. The leaves loose turgidity finally and complete drying of vines were observed. Examination of buried stem and roots of wilted vines show rotting and shredding of tissues of stem. The shredded portion is brownish and sticky. Some times sclerotia of *Sclerotium rolfsii* are visible on dead tissues. In some vines, the roots show terminal or interracalary knots or galls of nematode.

Pathogen

The disease is a complex. Following pathogens *viz. Rizoctonia solani, Sclerotium rolfsii, Phytophthora spp* and *Meloidogyne sp*. are associated with wilted plants.

Management

Sandy loam well drained soils should be preferred for cultivation of betelvine. Broad based furrow system of cultivation ensures better drainage and less incidence of wilt. Drip irrigation method reduces the incidence of wilt over flood or ridges and furrows system of irrigation. Soil Solarization in the month of May reduce the population of pathogens in soil.

1. Resistant Varieties

Varieties like Ramtek Bangla, Kalapakku, Maghai, and Meetha are tolerant to wilt under field conditions.

2. Alteration of Crop Practices

Crop rotation is very useful in reducing the inoculum of pathogens in soil. Intercrop of merigold reduces nematode population in soil. The support and shed plants like Pangara or Sheori is attacked by stem borer and the plants are dead. Such dead plants are harboured by soil inhabitant fungi like *Rhizoctonia solani and Sclerotium rolfsii* from which it spreads to betelvine. Hence instead of support and shade plants shading Net and Nylon ropes can be used to support the vine.

3. Biological Control

Application of neem seed cake reduces the population of pathogens in soil. *Trichoderma viridae* and *Trichoderma harziamum* preparations mixed with F.Y.M. can be applied for the biological control of soil borne fungi. Similarly for nematode control *Paeceilomyces sp* can be effectively used.

3. Chemical Control

1. Drenching of soil with 1 per cent Bordeaux mixture or 0.25 per cent copper oxychloride once in a month
2. Dipping of setts in 1 per cent Bordeaux mixture or copper oxychloride 0.25 per cent at the time of planting
3. Spray of 1 per cent Bordeaux mixture or copper oxychloride 0.25 per cent on leaves 2-3 times at one month interval during rainy season.
4. Application of granular insecticide like carbofuron @ 1 kg/ha after planting and leaves should not be harvested at least one month after insecticide application.

4. Integrated Management

1. Use of tolerant varieties
2. Soil solarization before planting
3. Crop rotation
4. Soil application of 1 per cent Bordeaux mixture or *Trichoderma* preparation @ 5 kg/ha
5. Dipping of cuttings in the solution of 1 per cent Bordeaux mixture or *Trichoderma* preparation.
6. Soil application of Namaticide or insecticide carbofuron for nematode control. Merigold should be intercropped with betelvine.
7. Use of shedding net and nylon rope for growing betelvine.
8. Use of broad based furrow and drip irrigation system for cultivation of betelvine.

Foot Rot of Black Pepper (*Piper nigrum* L.)

Y.R. Sarma

Indian Institute of Spices Research, Calicut

Symptoms

The disease consists of aerial phase and soil phase. Dark brown fast expanding lesions (spots) with fimbriate margins are noticed on tender and medium mature leaves. In addition tender green stems and runner shoots also are infected. Root infection starts from the feeder roots gradually spreads the larger roots and reach the

Foot Rot of Black Pepper

Leaf Infection

collar or foot (stem at the ground level) causing rotting. Independent collar infection is also noticed. The net result is complete death of the plant with in a span of 15-30 days and hence farmers call it as 'Quick wilt'.

Pathogen

Phytophthora capsici

Management

1. Use of Resistant Varieties

There is no high degree of resistance available among cultivated type. However tolerant lines like IISR-Sakthi, Balankotta, Uthirankotta and Narayakkodi have been identified. *Piper colubrinum*, a South American type showed multiple resistance to three major pathogens, *P. capsici, R. similis* and *M. incognita*. Incorporation of resistance of *Piper colubrinum* into *Piper nigrum* is attempted through biotechnological approaches. Pepper with root stock of *P. colubrinum* is successful only in areas where water table is consistently present.

2. Alteration of Crop Practices

Phytosanitation consisting of removal of infected and dead vines from the plantation, drenching of the spot with 5 litre of Copper oxy chloride (0.2 per cent) is recommended. Provision of good drainage, minimum tillage to avoid root damage during farm operations, shade regulation by lopping off branches of live standards during May are recommended to ensure better light penetration and better air circulation which are not conducive for high humidity build up and consequent less disease. Pruning off runner shoots arising from the base of the vines is essential to avoid foliar infections.

3. Biological Control

Trichoderma harzianum and *Pseudomonas fluorescens* have been identified as potential antagonists of *P. capsici*. Inoculum of *T. harzianum* raised in broken sorghum meal/coffee pulp/spent tea/talcum powder is mixed with nursery mixture used for raising fresh disease free rooted cuttings. *Glomous fasciculatum* (VAM) inoculum raised on maize roots is also mixed with nursery mixture and this protects the rooted cuttings from infection from *P. capsici, R. similis* and *M. incognita*. The inoculum is applied in the field @ 50gm (10^6 and above cfu) per vine mixed with 1kg of neem cake as a pre-monsoon treatment. Another 50gm of inoculum is applied during August-September as a post-monsoon treatment. Spraying and drenching the base of the vines with *P. fluorescens* inoculum also was found effective in reducing the disease (both foliar and root infection).

4. Chemical Control

Pre-monsoon spray during May-June and Post-monsoon spray during August-September with 1 per cent Bordeaux mixture is recommended to check foliar infections. Soil drenching around the base of the vines with 0.2 per cent Copper oxy chloride is also recommended both as pre-monsoon and post-monsoon treatments to check root infection. In place of Bordeaux mixture, Metalaxyl Mancozeb (Ridomil MZ72 WP) @ 1.25gm per litre of water or potassium phosphonate (Phytophos) @ 5ml per litre has been recommended both as a foliar spray as well as soil drench. Being systemic in nature these were found highly effective in checking the disease.

5. Integrated Management

This consists of raising of healthy rooted cuttings in nursery mixture either solarised or non-solarised but mixed with biocontrol inoculum as mentioned above is a major component since root infection starts right at the nursery but goes unnoticed without any visible symptoms. Phytosanitation coupled with application of biocontrol inoculum alongwith organics like neem cake etc. is the practice adopted. Potassium phosphonate as well as Metalaxyl Mancozeb are compatible with biocontrol agents and hence can be applied simultaneously. However, Copper oxy chloride is not compatible with *Trichoderma* and *P. fluorescens*. It is advised to give a gap of 15-30 days between the application of copper fungicides like Copper oxy chloride and biocontrol inoculum. Above all phytosanitation consisting of removal of dead and infected vines is absolutely essential to reduce the disease.

Capsule Rot of Small Cardamom

Y.R. Sarma

Indian Institute of Spices Research, Calicut

Symptoms

The infection starts on panicles spreading on the ground at the base of the clump and also on green immature capsules. It starts as water soaked spots and latter gradually spreads resulting in rotting of capsules as well as drying up of spikes. Disease starts during South-west monsoon (June-August) period. Rotting of leaves is noticed during December-January period. Rotting starts as water soaked spots and gradually spreads to entire lamina ultimately leading to shredding of leaves.

Pathogen

Phytophthora meadii/P. nicotiane

Capsule Rot of Small Cardamom

Management

1. Use of Resistant Varieties

Three important major varieties are Malabar, Mysore and Vazhuka are found to be susceptible and so far no resistance source has been identified.

2. Alteration of Crop Practices

Strict phytosanitory regulations involving removal of dead and infected clumps/capsule/spikes is essential to reduce the pathogen population buildup in the plantation. Heavy shade is highly conducive for the disease and hence lopping of the branches of the shade trees is practiced to reduce the shade and is generally done during May. Provision of drainage is important to avoid high soil moisture build up.

3. Biological Control

Trichoderma harzianum highly antagonistic to *P. meadii* has been isolated multiplied on coffee pulp/tea waste. The inoculum is applied at the base of the clumps as 50-100gm per clump (cfu 10^6 and above). This is applied both as a pre-monsoon (May-June) and post-monsoon (August-September) treatment.

4. Chemical Control

Pre-monsoon (May-June) and post-monsoon (August-September) spray and drench with 1 per cent Bordeaux mixture was found to be effective in reducing the disease. Soil drenching at the base of the clump with 1 per cent Bordeaux mixture is recommended. In place of Bordeaux mixture Potassium phosphonate is recommended (Akomin/Phytophos) and this is compatible with *Trichoderma harzianum*, the biocontrol agent.

5. Integrated Management

Integrated management involving phytosanitory measures application of *Trichoderma harzianum* along with Potassium phosphonate would be an ideal strategy to reduce the inoculum build up and consequent reduction in the disease. This should be combined with shade regulation and through provision of better drainage.

Chickpea

Wilt of Chickpea

O.D. Kohire

Marathwada Agricultural University, ARS, Badnapur, Dist. Jalna, M.S.

Symptoms

Yellowing and drying of leaves from the base upward, drooping of petioles and rachis, improper branching, withering of plants, browning of vascular bundles, and finally wilting of plants. Generally the disease occurs at two stages of plant growth (1) seedling stage, and (2) flowering or adult stage. The pathogen infects the root system by penetrating the epidermis, cortex and finding its way into xylem vessels.

Pathogen

Fusarium oxysporum f. sp. *ciceri*

Management

1. Use Resistant Variety

ICCV 2, JG 315, BGD 72, BG 244, Phule G 5, Pusa 391, ICCV 10, Gujrat Gram 1, JG 315, Avrodhi, JG 74, KWR 108, Vijay, JG 11, Annegiri, ICPH 2671.

2. Alternation of Crop Practice

Intercropping or crop rotation with crop like wheat, barley, linseed and mustard may help in management of wilt.

3. Biological Control

Use of bioagent like *Trichoderma viride, Trichoderma herzianum, Gliocladium virens, Bacillus subtilis, Pseudomonas fluorescens*, etc. in the management of wilt and other soil borne diseases in chickpea.

4. Chemical Control

Treating seeds before sowing with Benlate T (30 per cent Benomyl + 30 per cent Thiram) @ 1.5 kg/ha and Bavistin @ 2.5 g/kg of seed or Bavistin + Thiram (1+2 g/kg seed).

5. Integrated Management

Deep summer ploughing, resistant varieties, seed treatment, sowing date and spacing should be normal. Intercropping with linseed (2:1).

Dry Root Rot of Chickpea

O.D. Kohire

Marathwada Agricultural University, ARS, Badnapur, Dist. Jalna, M.S.

Symptoms

Yellowing and sudden drying of the plants. Normally, the leaves of the affected plants may be straw coloured and occasionally the lower leaves and stem may be brown. The tap root becomes dark brown, quite brittle in dry soil and often shows extensive rotting resulting in the loss of lateral roots. The fungus produces dark brown or black sclerotia on infected dead tissue, if the plants are pulled from the soil and examined, the basal stem and the main roots may show dry rot symptoms.

Pathogen

Rhizoctonia bataticola.

Management

1. Use of Resistant Variety

JG 315, K 850, DCP 92-3, Avrodhi, H 208 and C 104, ICC 12237 and ICC 12269.

2. Alteration of Crop Practices

Growing chickpea as intercrop/mixture with linseed or coriander effectively control the disease.

3. Biological Control

Seed treatment with *T. viride* + Vitavax @ 4:1 g per kg seed.

4. Chemical Control

Seed treatment with Captan or Thiram or PCNB @ 2.0 – 2.5 g per kg seed eradicates the seed borne fungus.

5. Integrated Management

Timely sowing early maturing varieties are recommended to escape the hot weather. Deep summer ploughing, use resistant varieties, intercropping with linseed or coriander.

Ascochyta Blight of Chickpea

O.D. Kohire

Marathwada Agricultural University, ARS, Badnapur, Dist. Jalna, M.S.

Symptoms

Initial circular spots appear near the tip of the young shoots and top most leaves. Later on brown spot appear in other aerial parts such as leaves, petioles, flowers, pods, branches and stems. The apical twig, branches and stems often show girdling, lesions on pods are prominent and usually circular with dark margins.

Pathogen

Ascochyta rabiei

Management

1. Use Resistant Varieties

Genotypes with moderate resistance are Gaurav (H 75-35), Pusa 244, Pusa 372, GNG 146, GNG 469, ILC 72, ICC 1069 and MCK 54.

2. Alteration of Crop Practice

Seed should be produced in disease free areas to prevent the introduction of disease in new areas or more virulent races into already infested areas. Crop rotation with wheat, barley, rapeseed and mustard particularly in the areas where the pathogen produce perfect stage.

3. Chemical Control

Seed treatment with Thiram, Benomyl, Calixin, Thiobendazole, Bavistin + Thiram (1:2), Captan, Captaf @ 3 g/kg seed and Rovral 2.5 g per kg of seed.

4. Integrated Management

Wide row and plant spacing and growing chickpea with wheat, barley or mustard. Seed treatment, and use resistant varieties, if cloudy weather prevails one protective spray of the fungicide should be applied.

Botrytis Gray Mould of Chickpea

O.D. Kohire

Marathwada Agricultural University, ARS, Badnapur, Dist. Jalna, M.S.

Symptoms

Initial Symptoms are water soaking and softening of affected plant parts *viz.*, leaves, flowers and tender shoots. On the stem, the gray mould symptoms are gradually replaced by dark gray and black sporodochia under humid condition, all the flowers are attacked resulting in the complete failure of the crop. BGM affected pod either do not produce any seed or produce only small-shriveled seed.

Pathogen

Botrytis cinerea.

Management

1. Use of Varieties

Varieties Pusa 209, Saki 9516 and Sweta should be used with other management practices.

2. Alteration of Crop Practice

Chickpea crop may be sown late in the first fortnight of November particularly in irrigated areas, lower seed rate should be used wide row and plant spacing should be adopted, tall erect and compact genotypes may be grown.

3. Biological Control

The application of *Trichoderma viride* was found equally effective as that of Vinclozolin (0.2 per cent).

4. Chemical Control

Seed treatment with Bavistin + thiram (1:2), bavistin @ 3 g/kg seed, thiobendazole @ 2 g/kg seed. Rovral or Ronilan @ 4 g/kg seed for elimination of seed borne infections of chickpea. Foliar sprays of Bayleton @ 200 g/acre in 100 liters of water for the control of secondary infection.

5. Integrated Management

Use resistant varieties; seed treatment with fungicides, wide row and plant spacing should be adopted.

Chilli

Virus Diseases of Chilli

Satya Prakash

I.A.R.I., Regional Station, Agricultural College Estate, P.O. Shivajinagar, Pune – 5

Chilli Mosaic

Symptoms

Characteristic symptoms are mosaic, mottling, oak-leaf pattern, vein clearing, vein banding, blistering, leaf distortion, streaks on stem and stunted growth (Figures 1, 2, 3). Fruits are few, deformed, smaller and generally show mosaic-mottling (Figures 4, 5).

Figure 1: Chilli Leaves Showing Mosaic, Mottling Due to TMV

Figure 2: Capsicum Plant Showing Mosaic, Leaf Distortion and Growth Stunting Due to CMV

Figure 3: PVY Infected Capsicum Leaf Showing Mosaic and Oak-leaf Pattern

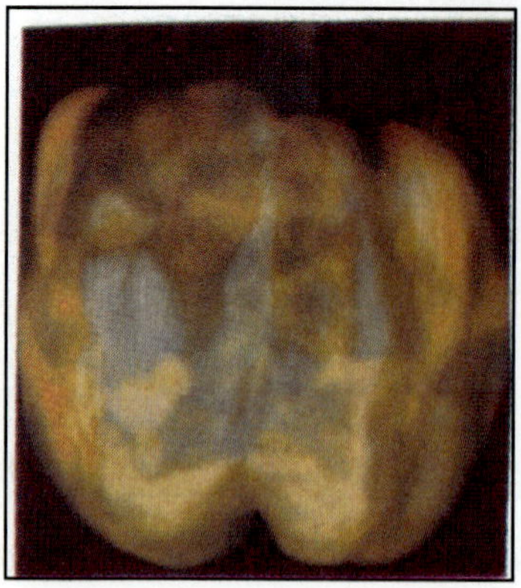

Figure 4: CMV Infected Fruit of Capsicum Showing Mosaic

Figure 5: Capsicum Fruits Showing Deformations Due to CVMV

Figure 6: Capsicum Leaves Exhibiting Veinal Necrosis Due to TSWV Infection

Pathogen

Potato X Potexvirus (PVX); potato Y (PVY), pepper veinal mottle (PVMV), pepper vein banding (PVBV), tobacco etch (TEV) and chilli veinal mottle (CVMV) Potyviruses; tobacco mosaic Tobamovirus (TMV), and cucumber mosaic Cucumovirus.

Chilli Necrosis and Die-back

Symptoms

Concentric zones on leaves, mosaic mottling on terminal leaves with dense coalescence of small rings, spots and concentric rings, veinal necrosis (Figure 6), necrotic streaks on stem, sudden yellowing and browning, necrotic spots, necrosis of terminal shoots and die-back symptoms. Fruits exhibit chlorotic spots, green and red areas surrounded by yellow halos, large necrotic blotches and at times concentric rings, necrosis and abortion of flowers are common.

Pathogen

Tomato spotted wilt Tospovirus (TSWV) and tobacco streak Ilarvirus (TSV).

Chilli Leaf Curl

Symptoms

The characteristic symptoms of the virus infection are: vein clearing, curling, blistering, puckering, and heavy crinkling of the leaves with chlorosis and enations. Infected plants are stunted and often bushy in appearance.

Pathogen

Tobacco leaf curl Geminivirus.

Management

1. Selection of seeds from healthy plants.

2. Disinfection of seeds with Trisodium phosphate (TSP) (0.3 per cent) for 48 h in order to eliminate the externally borne tobacco mosaic virus (TMV) from the seeds.
3. Nursery should be grown under white nylon net.
4. Spray the nursery bed with monocrotophos or metasystox (0.1 per cent) at 15 days intervals to control insect vectors.
5. Soil application of carbofuran or phoret or thymet-10 g @ 1.5 kg a.i./hectare at the time of transplantation of seedlings in the field.
6. Sowing barrier crops (6-8 lines) like maize, jowar, bajra or sorghum around chilli/capsicum plot at least 60 days in advance.
7. Use of silver plastic mulch in field to create confusion to the landing site of the insects, since it reflects UV light that repels the insects.
8. Yellow sticky polythene sheets erected on the windward side of chilli/capsicum field helps to reduce virus incidence by trapping the insect vectors.
9. Spray of two insecticides (metasystox and monocrotophos-0.1 per cent) alternately in the field at 15 days intervals.
10. Remove volunteer plants emerging at the edges of a planting or within a field since they may be potential virus carriers.
11. Remove and rogue infected plants from the field as soon as they are noticed.

Fruit Rot of Chilli

V.V. Sonawane

Department of Botany, Vasantrao Naik Mahavidyalaya, Aurangabad – 431 003

Symptoms

Small, black, circular lesions appear on the skin of the fruits and expend towards the long axis giving ultimately elliptical shape. The infected portions form dark depressions, as the infection advances, the spots turn either diffused and black, greenish or dirty – grey in colour, or they are markedly delimited by a thick and clear black outline surrounded by a light black or light brown – coloured spot. On the infected areas, numerous black dots like acervuli of the fungus may be formed. In more advanced stages, seeds inside the fruit becomes turn rusty. The attacked parts turn dark and become depressed or wrinkled, and ultimately the fruit shrivels and dries up.

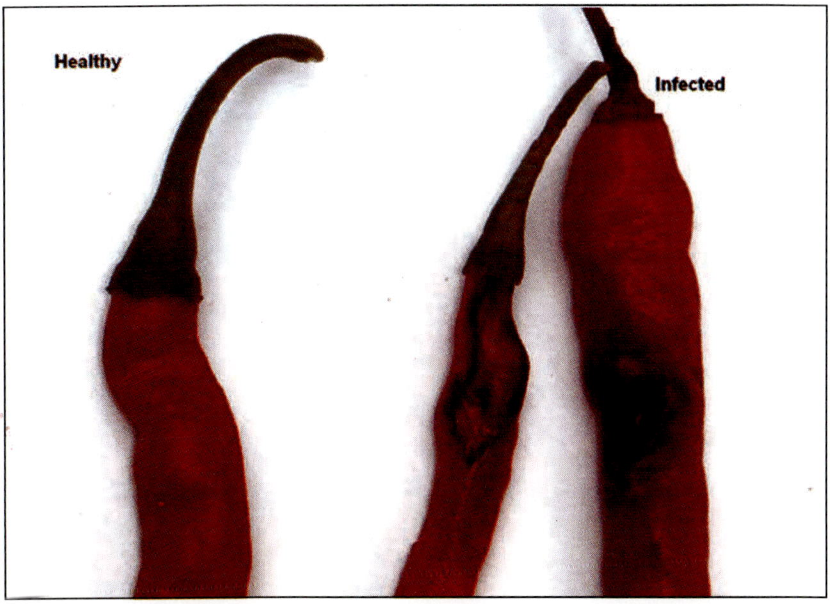

Pathogen

Colletotrichum capsici (Syd.) Butler and Bisby

Management

1. Spraying captafol @ 0.2 per cent followed by copper oxychloride @ 0.25 per cent and carbendazim @ 0.1 per cent.
2. Biocontrol by spraying 1000 ppm plant extracts of *Acorus calamus* and *Piper betel.*
3. New variety MI - Hot 7 is identified as moderately resistant to fruit rot.
4. Variety Pant C-1 is resistant.
5. Check the secondary infections by spraying copper fungicides.

Powdery Mildew of Chilli

M.P. Kulthe

Department of Botany, Vasantrao Naik Mahavidyalaya, Aurangabad – 431 003

Symptoms

The disease appears as small, circular, whitish spot in both the surfaces of the leaves. In later stages the entire leaf surface is covered by dirty-white growth. The infection starts from the older leaves but gradually covers the entire plant. In such cases, the lamina develops complete or partial chlorosis and ultimately plant dries up. With age, the mildew mass turn grey and develops numerous minute black bodies distributed all over the surface.

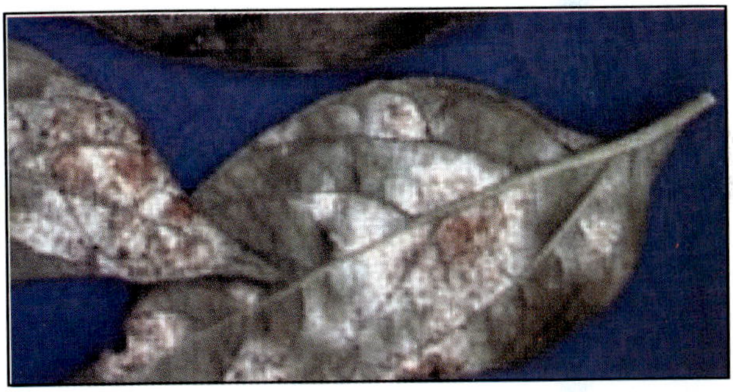

Pathogen

Leveillula taurica (Lev.) Arn.

Management

1. The best control was achieved with the application of penconazole (PDI of 9.03), followed by propiconazole (PDI of 13.84), the highest yields of 7.97 and 7.40 q/ha, respectively. However, the cost:benefit ratio of carbendazim (1:8.44) and wettable sulfur (1:8.43) were considered as the most economical.
2. Application of hexaconazole 0.05 per cent, wettable sulfur 0.25 per cent, carbendazim 0.10 per cent, tridemeton 0.05 per cent and thiophanate-methyl 0.05 per cent, triadimefon 0.05 per cent were the most effective in reducing

the disease. The percentage disease control of the fungicides ranged between 62.40 and 73.60 per cent.

3. Spray thrice at 10 – 15 days interval with 1 ml Dinocap or 1 gm wettable sulphur/litre water.
4. Before flowering, dust the 8 – 10 kg sulphur powder/acre is also very useful.
5. Use resistant varieties like Hisar Shakti, Hisar Vijay, Jawahar Mirch.

Leaf Spot of Chilli

D.A. Doiphode

Department of Botany, Vasantrao Naik Mahavidyalaya, Aurangabad – 431 003

Symptoms

Leaf spots are 3-6 mm in diameter; first appear as water – soaked areas, which later have white center with dark – brown margins. They are usually oblong or circular. Severely spotted leaves turn yellow and drop. Spotting of the leaf petioles also is a factor in the defoliation of the plants. Loss of leaves is often so serious as to cause drastic reduction in both size and quality of the fruit.

Pathogen

Cercospora capsici Heald and Wolf

Management

1. Rotation of crops, good drainage and spraying with fungicides.
2. Use of disease – free seeds
3. The spray of fungicides are Bordeaux Mixture (3:3:50), Ziram 2 per cent, Captan 0.3 per cent or Maneb 0.2 per cent. Also spray twice at 10 – 15 days interval with 2.5 gm Mancozeb or 1 gm carbendazim/liter of water.
4. Treat the seeds with copper sulphate (32 gm in 16 litres of water).
5. Use moderately tolerant varieties like MI – Hot 7

Citrus

Citrus Greening

D.K. Ghosh

NRCC, Nagpur- 440010, Maharashtra

Symptoms

Greening infected leaves were small; upright and frequently have various types of chlorotic mottling and Zn-deficiency like symptoms. In severe cases leaves become almost chlorotic with scattered dark green islands. The internodes become shorter, plants are very stunted and many of them never yield any fruit. Symptoms are often seen on part of the canopy or even only in a branch or twig. Chronically infected trees are sparsely foliated and show extensive twig die-back. Infected fruits are small, misshapen, may fall prematurely and contain aborted seeds.

Incidence of greening is more on sweet orange as compared to Nagpur mandarin, Coorg mandarin and acid lime.

Leaf Mottling Due to Greening

Pathogen

Disease is caused by a phloem-limited, uncultured bacterium *Candidatus liberibacter* asiaticus and transmitted by vegetative propagation and psyllid vector, (*Diaphorina citri* Kuwayama). Under experimental conditions the pathogen is transmitted successfully from sweet orange seedlings to periwinkle (*Catharanthus roseus*) by dodder (*Cuscuta* sp.).

Diagnostic technique for the greening pathogen includes biodiagnosis in sweet orange indicator plants, TLC, PCR (polymerase chain reaction), DNA-DNA hybridization using nucleic acid probes etc.

Management

Integrated management approach includes removal of infected unproductive trees and their replacement by disease free plants. Since the disease is spread through psylla, suitable insecticide eg Acephate @1.25 g/l or Imidacloprid @ 4 ml/10 liter or Abamectin @ 3 ml/10 liter of water should be sprayed to control the insect vector. Spraying should be done at bud burst stage or when infestation is noticed on sprouts in June-July, Jan-Feb and Oct-Nov. Spraying with Ledermycin 600 ppm @ 10 liter per plant (to be dissolved first in alcohol followed by mixing in water) may suppress the disease symptom. Application of mixture of $ZnSO_4$ and $FeSo_4$ @ 50-100g/tree/year each in two split doses (July and Sept) helps in greening management. Disinfections of pruning tools with 1-2 per cent sodium hypochlorite solution prevent spread of the disease by mechanical means. Murrya plants should be removed which otherwise helps Psylla population for survival and multiplication.

Black Pod of Cocoa

M. Gunasekaran and R. Chandramohanan

Central Plantation Crops Research Institute, Kasaragod – 671 124, Kerala

Symptoms

1. The disease is common during rainy season and reduces the yield.
2. Infection appears as a chocolate brown spot on fruits of all ages (Figure 1), which later spreads and covers the entire surface of the pods.
3. As the disease advances, whitish growth of the fungus is seen on the affected pods (Figure 2). Later the pods turn black. The internal tissues as well as beans are discoloured.

Figure 1

Figure 2

Pathogen

Phytophthora palmivora

Management

1. The disease can be controlled and prevented by spraying 1 per cent Bordeaux mixture at the onset of southwest monsoon and thereafter at least twice during the monsoon season at an interval of 45 days.
2. Frequent removal and destruction of infected pods should be carried out regularly.
3. Over crowding of branches should be avoided by regular pruning.

Canker of Cocoa

M. Gunasekaran and R. Chandramohanan

Central Plantation Crops Research Institute,
Kasaragod – 671 124, Kerala

Symptoms

1. The canker appears either on the main trunk (Figure 1 and 2) or fan branches.
2. Initially greyish brown water soaked lesions appear on the bark. Later, reddish brown liquid oozes from these lesions, which dry up to form rusty deposits.
3. The tissues beneath the lesions show reddish brown discoloration due to rotting.
4. When the cankers girdle the main stem or branches, die back symptoms appear and in severe cases the tree may ultimately die.

Pathogen

Phytophthora palmivora

Figure 1

Figure 2

Management

1. Remove the infected tissues and apply Bordeaux paste (10 per cent).
2. All black pod infected fruits should be removed and destroyed.
3. Provide good drainage in the garden.

Charcoal Pod Rot of Cocoa

M. Gunasekaran and R. Chandramohanan

Central Plantation Crops Research Institute,
Kasaragod – 671 124, Kerala

Symptoms

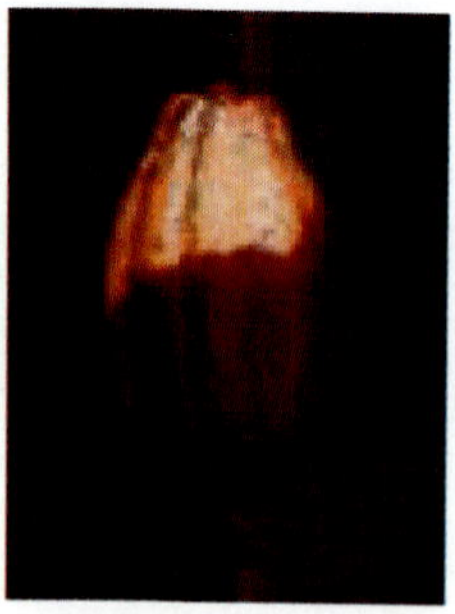

1. Infection appears as black or brown spot.
2. Affected pods turn black and remain on the tree as mummified fruits.
3. Internal tissues are rotten and the beans turn black.
4. This disease is known to occur throughout the year but severe during summer.
5. Pods of all ages are susceptible.

Pathogen

Botryodiplodia theobromae

Management

1. Spraying 1 per cent Bordeaux mixture to fruits is recommended to control the disease.
2. Rodent and insect damage should also be controlled.

Vascular Streak Die Back of Cocoa

M. Gunasekaran and R. Chandramohanan

Central Plantation Crops Research Institute, Kasaragod – 671 124, Kerala

Symptoms

The first visible symptom is yellowing of a single leaf in the second or third flushes from the tip of twigs with characteristic green islets. Such infected leaves fall prematurely and leaves above and below to it also develop similar symptoms.The brown streaking of the wood originating from the scars of fallen leaves coupled with the presence of the fungus in xylem tissues. Finally the whole plant dries even though most of the affected leaves remain attached to the plant.

The disease is reported from Kerala and recently in a severe form from Karnataka.The disease has become major impediment in establishing cocoa gardens in Mysore district.

Pathogen

Oncobasidium theobromae

Management

Regular pruning of the infected branches 30 cm below the last spot of visible streak and stem painting with propiconazole helps in reducing the spread of the disease.

Coconut

Bud Rot of Coconut

M. Gunasekaran and R. Chandramohanan

Central Plantation Crops Research Institute, Kasaragod – 671 124, Kerala

Symptoms

The earliest visible symptom is the withering of the spindle marked by pale colour. In the adult palm the spindle (spear leaf) turns dull and bends over. In seedlings, the spindle turns brown and comes off easily with a gentle pull. The base of the spindle rots and emits a foul smell. If allowed to degenerate further, the leaves next to the spindle also begin to rot leaving only the fully matured leaves on the crown (Figure 1). In the end, the palm succumbs to the disease. Simultaneously, the infection proceeds to the deeper softer tissues of the cabbage and finally attacks the growing

Figure 1

tip. Once the apical meristem is affected, the palm dies. Symptoms of attack of the fungus can be noticed as water-soaked lesions at the base of the petiole of the leaves also. The older nuts persist on the crown for some time, while the younger ones may fall off.

Pathogen

Phytophthora palmivora (Figure 2 A and B. The Sporangia)

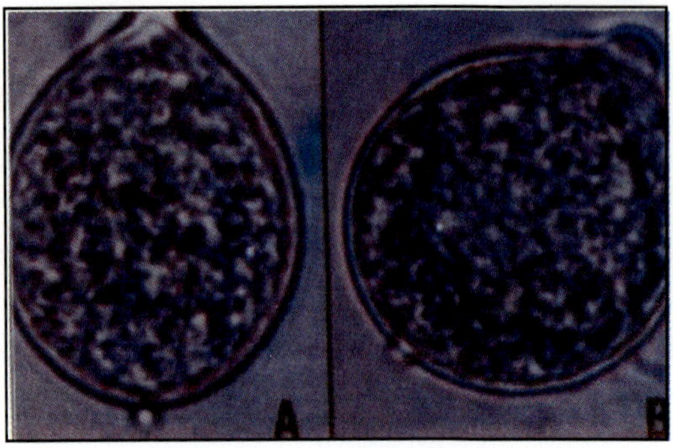

Figure 2A&B

Management

1. Regular spraying with any copper fungicides, *viz.*, 1 per cent Bordeaux mixture, Dithane M-45, or copper oxychloride at 40 days interval especially before and after monsoon is an effective preventive measure.
2. Perforated sachets containing 3g Dithane M-45 should be kept in leaf axils of copper sensitive palms during rainy season.
3. Cut and remove severely infected or dead palms to control secondary spread of the disease as well as to reduce the inoculums in the garden. Rotten portions of the crown should be destroyed by burning.
4. In the early stage of disease incidence, cut and remove all infected tissues. After thorough cleaning of the cut surface, apply 10 per cent Bordeaux paste. The treated portion should be covered with polythene sheet or mud pot.
5. In adult palms, it is possible to keep bunches until the nuts are harvested. However, treat the infected bud region with Bordeaux mixture or copper fungicide to kill the mould.
6. As a prophylactic measure, all the healthy palms surrounding the diseased ones should be sprayed with 1 per cent Bordeaux mixture on the spindle and base of 2 – 3 inner most leaves.
7. Systemic fungicides can also be applied as prophylactic before the onset of monsoon so that they can protect the palms during the critical period.
8. Stem injection or root feeding of systemic fungicides, Akomin (16.8ml), Aureofungin – sol (36.4 g/palm) or Calixin (21 ml/palm), is effective in protecting the palms for a period of eight weeks after application.
9. Plant resistant/tolerant cultivars with proper spacing. The low lands with generally high humidity are highly favourable for the development of the disease, especially when the drainage is poor.

10. Crown cleaning is strictly advised as *Phytophthora* propagules survive in the crown debris of both diseased and healthy palms in the off-season. This is a potential inoculum that can serve as the source for the subsequent season.
11. Sequential application of botanicals and biocontrol agents, *T. harzianum* and *B. amyloliquefaciens*, with suitable interval can effectively suppress the disease.

Basal Stem Rot of Coconut

M. Gunasekaran and R. Chandramohanan

Central Plantation Crops Research Institute, Kasaragod – 671 124, Kerala

Symptoms

Withering followed by yellowing and drooping of older leaves in a quick succession is the characteristic initial visible symptom (Figure 1). Even before this, decay of finer roots would have started. The root decay gradually extends up to the bole region and from there to the stem. This is accompanied by exudation of reddish brown viscous fluid from the basal portion of the trunk (Figure 2). The bleeding patch enlarges in size and traverses upwards, to a height of 3 to 4 metres in certain cases. There is commensurate internal decay just below the bleeding patch. More leaves droop and bunches subtended by these leaves also droop, often resulting in nut fall. The crown dries, weakens and topples over in the slightest wind. Dead palms and

Figure 1

Figure 2

stumps show the presence of brackets of *Ganoderma* (Figure 3). The disease progresses rapidly in dry areas during summer especially in sandy soils.

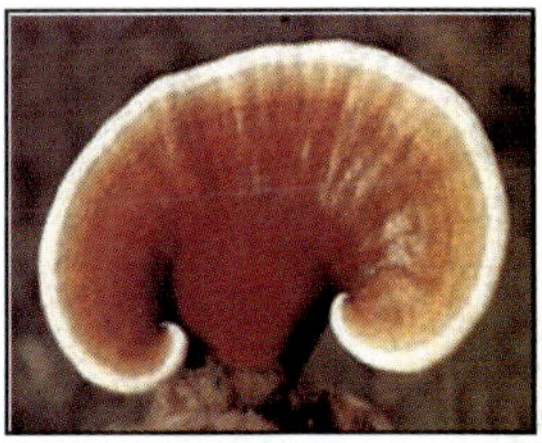

Figure 3

Pathogen

Ganoderma applanatum and *G. lucidum* (Leys) Karst.

Management

1. Phytosanitation

Palms in the advanced stage of disease and giving uneconomic returns should be cut and removed and the root system along with the bole burnt. The bleeding patch in the stem may be chiseled and protected with Tridemorph (5 per cent Calixin) and subsequently with hot coal tar.

2. Isolation Trenches

The palms in the early or middle stages of the disease should be isolated from the neighbouring healthy palms by taking trenches of 1M deep and 30 cm wide to prevent root contact.

3. Cultural Practices

Avoid repeated ploughing, closer spacing, flood irrigation, and water logging. Grow *Ganoderma* resistant inter/mixed crops like banana in the garden.

4. Nutrient Application

Apply fertilizers @ 500:320:1200 g. NPK per palm per year and enough organics (at least 50 kg per palm). In addition, each palm should be given 5 kg neem-cake during September-October.

5. Chemical Treatment

The diseased palms may be treated with Aureofungin - sol @ 2.0g + 1g copper sulphate in 100 ml. water through root feeding thrice a year in July, October and January. Treatment with calixin 5 per cent at quarterly intervals was also found to be equally effective.

Whenever *Xyleborus* or *Diocalandra* attack is noticed on the stem of affected palms, swabbing with Carbaryl (Sevin 10 per cent) WP may be done.

6. Containing the Disease

Do not transport seedlings from infected area to healthy area. In sporadic cases of occurrence, the severely affected palms should be removed and other infected palms isolated from healthy ones by taking trenches in two circles. The palms in early stages of disease are to be treated with Aureofungin-sol as mentioned above. Each palm may be given 5 kg neem cake in addition to the regular fertilizer and recommended dose of organics.

Stem Bleeding of Coconut

M. Gunasekaran and R. Chandramohanan

Central Plantation Crops Research Institute, Kasaragod – 671 124, Kerala

Symptoms

The typical symptom of stem bleeding is the exudation of reddish brown gummy fluid from growth cracks on the trunk, which becomes black on drying (Figure 1). In the initial stages of infection, bleeding symptoms appears only one or two longitudinal cracks at the base and later spreads upward on the stem. The longitudinal patches may coalesce to form larger patches. The tissue inside the lesions shows discoloration and decay. In young palms under cooler conditions, the decay may extend into the deeper layers of central cylinder. As a result of extensive damage in the stem tissue, the outer whorl of the leaves turn yellow, dry and shed prematurely. The trunk tapers and reduces the crown size. In the advanced stages, scoletid beetles like *Diocalandra* and *Xyleborus* infest the palms and further weaken the stem.

Figure 1

Pathogen

Thielaviopsis paradoxa (Figure 2: The mycelium and conidia)

Management

1. The affected bark should be chipped off and paint with 5 per cent Calixin (5 ml Calixin in 95 ml water). The painted wound should be left to dry for one day and then covered with molten coal tar.

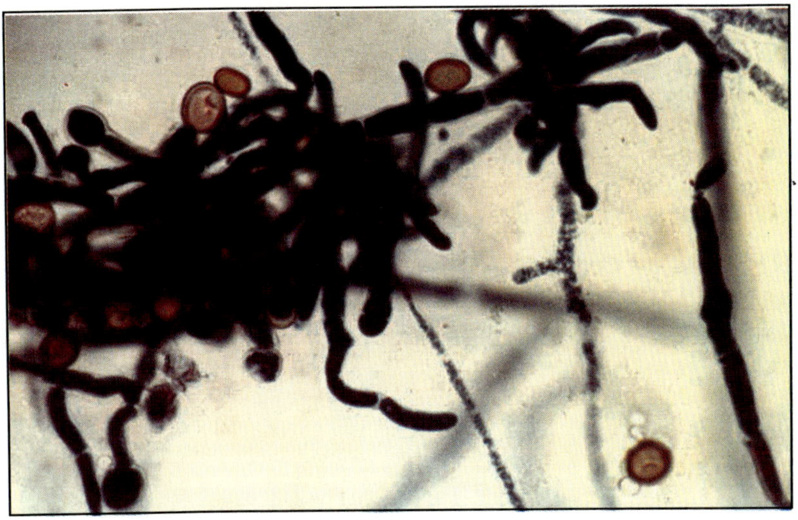

Figure 2

2. Root feeding of 5 per cent Calixin at quarterly intervals is effective in controlling the disease.
3. Apply neem cake @ 5 kg/palm/year in the basins during Sept – Oct.
4. Avoid wounding of coconut trunk.
5. Provide regular basin irrigation during summer months.
6. Good drainage facilities may be made in the garden, wherever necessary.

Wilt of Coconut

M. Gunasekaran and R. Chandramohanan

Central Plantation Crops Research Institute,
Kasaragod – 671 124, Kerala

Symptoms

The most obvious and diagnostic symptom of the disease is the abnormal inward bending of the leaflets termed ribbing or "flaccidity". Foliar yellowing of the outer whorl of leaves and marginal necrosis are the other associated symptoms (Figure 1). In seedlings and juvenile palms yellowing of foliage is virtually absent and flaccidity is the only symptom evident. Instances of diseased palms having their outer leaves quite healthy but with yellowing of leaves in some of the inner whorls are also noticed.

Figure 1

With the progress of the disease, extensive rotting of roots is observed. Shedding of immature nuts if the tree is bearing at the time of disease contraction is yet another symptom observed in some cases. Drying up of spathes and necrosis of spikelets from tip downward in unopened inflorescence is noticed in certain cases. The oil content is very much reduced and the oil losses its flavour as well.

Pathogen

Phytoplasma (Figure 2: The electron micrograph of Phytoplasma) are suspected as the cause of the disease

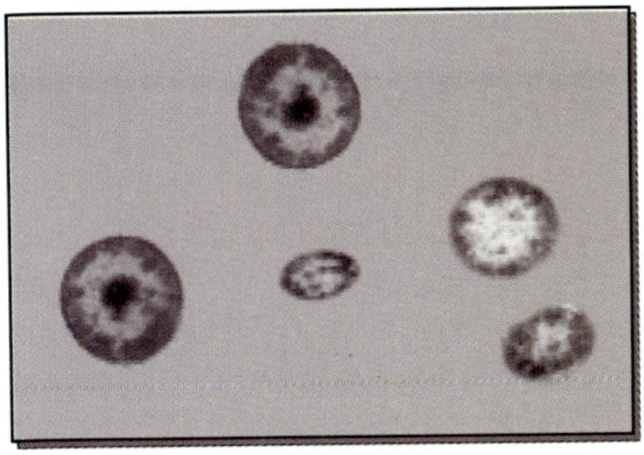

Figure 2

Disease Management

For the Heavily Diseased Contiguous Area

1. Remove all severely infected adult palms and palms in the prebearing age (juvenile palms).
2. Nutrient management: Balanced fertilizer application, addition of organic manures, raising of green manure crops in the basins and incorporation, irrigation during summer months, practicing plant protection methods and adopting inter and mixed cropping.
3. Mixed farming in the diseased gardens involving the raising of fodder crops in the inter spaces, maintaining milch cows and recycling of organic waste will increase the yield of palms and consequently the income from unit area.
4. Mixed cropping with cocoa and intercropping with tapioca, elephant foot yam and yams in the disease-affected gardens increase the mean nut yield and slow down the progression of the disease.
5. For the mildly affected area: Systematic rouging of diseased palms in the mildly affected area could prevent the further spread of the disease.
6. Provide irrigation in summer.
7. Adopt control measures for leaf rot disease. As leaf rot occurs super imposed on root (wilt) disease and reduces the yield of the palm drastically.

Leaf Rot of Coconut

M. Gunasekaran and R. Chandramohanan

Central Plantation Crops Research Institute, Kasaragod – 671 124, Kerala

Symptoms

Leaf rot occurs, superimposed on RWD palms, as minute, water soaked lesions on the emerging spindle with different shades of colour and shape. These lesions enlarge, coalesce freely leading to extensive rotting. The rotten portions dry up, turn black and fall off. Tips of leaflets and midribs often become blackish and shriveled. The progress of rotting slows down with the maturity of the leaflets. Continuous attack of newly emerging spindle leaf results in the gradual exhibition of similar symptoms in all the leaves in the crown (Figure 1). Sometimes the decayed leaflets glue together so that spindle does not open out. Though the disease does not kill the palm outright, its slow progress results in reduction of leaf surface in the crown, which causes steady decline in the yield. Palms of all ages are susceptible to the infection.

Figure 1

Pathogen

Colletotrichum gloeosporioides and *Exserohilum rostratum* (Figure 2.The Conidia of *E. rostratum*).

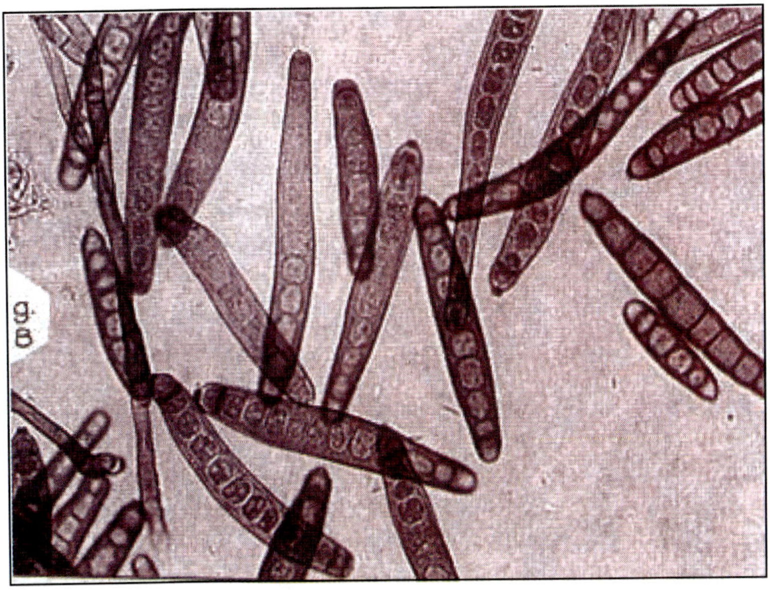

Figure 2

Management

Leaf rot can be effectively controlled by:

1. Cutting of the rotten portions of the spindle and the adjacent two innermost fully opened leaves.
2. Pouring either Contaf - 5EC 2 ml or Mancozeb (Indofil M-45) 3 g in 300 ml of water in the well around the base of the spindle leaf during April-May and October –November.
3. Application of Furadan 3G 30 g or Phorate 10G 20 g mixed in 200 g of sand around the base of the spindle leaf. Treatment of all the palms to be carried out twice a year *i.e.,* during April- May and Oct – Nov. along with harvest of nuts.

Crown Choke of Coconut

M. Gunasekaran and R. Chandramohanan

Central Plantation Crops Research Institute, Kasaragod – 671 124, Kerala

Symptoms

Emergence of shorter leaves with deformed and crinkled leaflets is the first symptom of the disease. Leaflets often show severe tip necrosis. Number of leaflets reduces gradually and hooking is also seen. Affected leaflets do not unfurl properly and in many cases give a choked appearance to the frond. As the disease progresses stick-like leafstalks emerge. The inner leaves crowd round the bud, and prevent normal unfurling of the flag leaf. Necrosis may proceed to the deeper tissues and kill the meristem. Premature nut falls is seen and gradual reduction in yield is noticed. Inflorescence emergence is hindered resulting in yield loss.

Cause: **Boron deficiency.**

Management

1. Soil application of Borax @ 50g/palm at half yearly intervals, during Feb-March and Aug-Sept., improves the condition of the palm.
2. Making slits in the tight extension of the petioles and opening out the crown would help in the normal emergence of fronds.

Gray Leaf Spot Disease

M. Gunasekaran and R. Chandramohanan

Central Plantation Crops Research Institute, Kasaragod – 671 124, Kerala

Symptoms

The symptoms appear on the outer whorl of leaves as minute yellow spots with a gray brown margin. The lesions may be oval in shape measuring up to 5 cm length. The center of the spots turns grayish white while the colour of the margin deepens. Many spots coalesce to form large irregular necrotic patches. The leaves in advanced stage of infection present blighted appearance. Complete drying and shriveling is common when the infection is severe. On the upper surface of the leaf the black pycnidia of the fungus appears as minute specks. The disease causes serious damage in nursery plants, but in adult palms the infection does not cause serious damage.

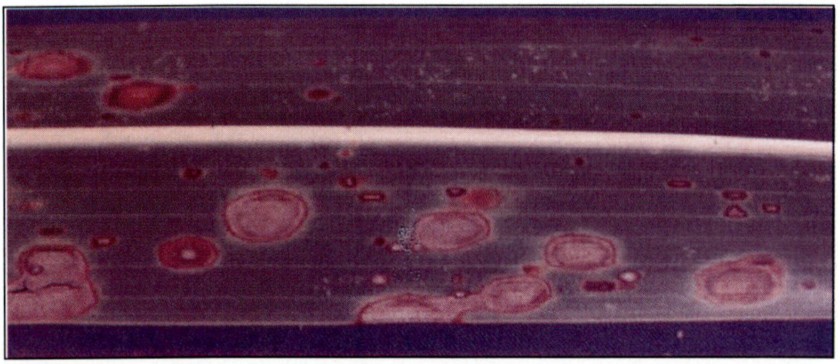

Pathogen

Pestalotiapsis palmarum

Management

1. Remove severely affected leaves and burn.
2. Spray the foliage with 1 per cent Bordeaux mixture or 0.3 per cent Mancozeb.
3. Apply potassic fertilizers
4. Improve the drainage of the gardens.

Nut Rot

M. Gunasekaran and R. Chandramohanan

Central Plantation Crops Research Institute, Kasaragod – 671 124, Kerala

Symptoms

1. Sporadic incidence of abnormal rotting/drying of immature nuts *in situ* was observed in Kerala state, India.
2. Infection usually originates from the apex. Early symptoms are dark grey to brown lesions with wavy to undulated margins starting from the apex of the nuts. Lesions progress covering a large area of the nuts.
3. Affected nuts become desiccated, shrunk, deformed and remained attached to bunch.
4. In some cases the contents leak owing to splitting of such nuts.
5. When cut open, the infected nuts reveal extensive rotting of the epicorp, which spread to meso and endocarps as well, especially in 6-8 month old nuts.
6. Symptoms caused by *Phytophthora* sp. slightly differ. Near the stalk, a discoloured area is developed, which will appear at first water-soaked and darker green than the rest of the surface of the nut. Later on, the lesions turn brownish in colour and appear as depressions due to the decay of the underlying tissues.

Pathogen

Phytophthora palmivora and *Lasiodiplodia theobromae*

Management

1. Spraying with 1 per cent Bordeaux mixture during the pre-monsoon and post monsoon periods can successfully control fruit rot due to *Phytophthora* infection.
2. For *Lasiodiplodia* infection, spraying 0. 1 per cent Bavistin is recommended.
3. Regular phytosanitory practices, thorough cleaning of the crown and collection of shed nuts debris and burning should be adopted.

Bud Rot of Coconut

P.A. Fugro and K.P. Vaidya

Department of Plant Pathology, Dr. Balasaheb Sawant Konkan Krishi Vidyapeeth, Dapoli – 415 712, M.S.

Symptoms

Bud rot of coconut is a continuous rainy season problem in coconut plantations of Konkan region (coastal belt) of Maharashtra.

The heartleaf of the tree mostly 12-45 years of age turns pale and brown. It bends over and breaks down. Some of the young leaves loose colour and droop before breaking down of heartleaf, which can be easily pulled out at later stage. Soon the leaves in surrounding whorl turn yellow to brown. The outer leaves and spathes around the central spindle falloff. Young nuts do not grow and fall down. The infected soft portion of the bud rots becomes slimy mass and emits a foul smell. Vegetative bud is killed and the tree succumbs to death.

Pathogen

The pathogen inciting bud rot is *Phytophthora palmivora* Butl.

Management

The only available measure to check the further spread of the disease is to cut down the affected and suspected crowns and burn them immediately. In early stages of infection, application of Bordeaux paste on crown after cleaning and removal of infected portion is advisable but very time consuming and laborious.

Healthy trees in the vicinity of the diseased in the plantation should be given prophlylactic sprays of 1 per cent Bordeaux mixture. Most of the attempts made for stem injections with systemic fungicides like Alliette or Metalyxil have not yielded fruitful results so far. Sanitation in the plantation should be given due attention. It's advisable to maintain proper drainage and maintain the spacing of 7.5 x 7.5 m between two plants.

Cotton

Bacterial Blight/Angular Leaf Spot of Cotton

P.K. Chakrabarty, R.L. Chavhan and Rita Dhanole

Central Institute for Cotton Research, PB No. 2, Shankarnagar, Nagpur – 440 010

Symptoms

Water-soaked, angular spots on lower surface of the leaves characterize disease. These spots turn black later. In severe cases, several spots may coalesce to form large dead areas resulting in seedling blight. The veins of leaves, stem, petioles and branches may develop elongated lesions. The infection may spread down the petioles and extend on the stem. It is at this stage the maximum loss occurs. The bolls are also often affected forming circular spots with greasy appearance under hot and humid conditions. Depending upon the parts of the plants affected, stages of infection or symptoms, the disease is referred variously as seedling infection, angular leaf spot, black arm and boll rot.

Angular Leaf Spot on Leaf

Seedling Blight

Bacterial Spot and Boll Rot

Pathogen

Xanthomonas axonopodis pv. *malvacearum.*

Management

1. Resistant Varieties

Cultivation of resistant varieties of cotton *viz.* BJR (Bacterial blight Jassid resistant), LH 900 and Arogya (*G. hirsutum*); H-420 and Virhar (*G. arboreum*); Digvijay and Jayawant (*G. herbaceum*)

2. Agronomic Practices

Removal and destruction of volunteer plants from the fields are very essential, as they are the chief source of infection.

3. Biological Control

Cotton phyllolplane bacteria namely, species of *Aeromonas, flavobacterium* and *Pseudomonas* on preinoculation may protect plants from infection.

4. Seed Treatment

Acid-delinting followed by seed dressing with carbendazim @3-4g/kg of seed eliminates seed-borne inoculum. Hot water treatment of seeds at 56°C for ten min eliminates both external and internal infection of pathogen.

5. Chemical Treatment

100 mg/ml of any indicated antibiotics (tetracycline, streptomycin sulphate, plantamycin, poshamycin) with 3 gm copper oxychloride per lit water at 15 days interval from the time of disease appearance.

Neem-based eco-friendly botanical pesticide like Agricare and Planotype are also effective.

Grey Mildew of Cotton

P.K. Chakrabarty, R.L. Chavhan and Rita Dhanole

Central Institute for Cotton Research, PB No. 2, Shankarnagar, Nagpur – 440 010

Symptoms

Disease initially appears on older leaves in the form of irregularly angular, translucent spots, with a definite and irregular margin delimited by the veins of the leaf. The lesions are light to yellowish green on the upper surface of the leaves. As the spots grow older, the leaf tissues turn yellowish brown while whitish frosty growth appears chiefly on the under surface but occasionally also on the upper surface. This is the conidial stage of the fungus. As the spots grow older, the leaf tissues turn brown. The affected leaves finally drop. The lesions also occur on the bracts subtending the bolls.

Grey Mildew (*Ramularia areola*)

Pathogen

Ramularia areola

Management

1. Resistant Varieties

Grow resistant varieties, eg. Sujata, Suvin, IC 629 and SB289.

2. Agronomic Practices

Avoid continuous cultivation of cotton. Destroy the infected-plant debris.

3. Chemical Treatment

Spray crop with Bavistin @ 2 g/l or wettable sulfur @ 3g/l, 2-3 times at weekly intervals from appearance of the disease. If prevalent the spray schedule can be repeated.

Root Rot of Cotton

P.K. Chakrabarty, R.L. Chavhan and Rita Dhanole

Central Institute for Cotton Research, PB No. 2, Shankarnagar, Nagpur – 440 010

Symptoms

Disease occurs in patches that enlarge year after year. It appears either in the seedling stage or after wood formation. In the young seedling a yellow patch appears on the lower part of the stem and if the attack is severe the seedling may collapse. In older seedlings patch becomes blackened at ground level and the seedlings dry up and wither. Affected plants can be pulled easily out of ground. With the exception of the tap-root and a few secondary roots, the others are all decayed. The tips of the roots are moist, discoloured yellow and sticky with soil adhering to them (Figures 1–3). The bark is shredded on the roots and basal part of the stem.

Figure 1: Field View of Root Rot

Figure 2: Damping Off of Seedlings

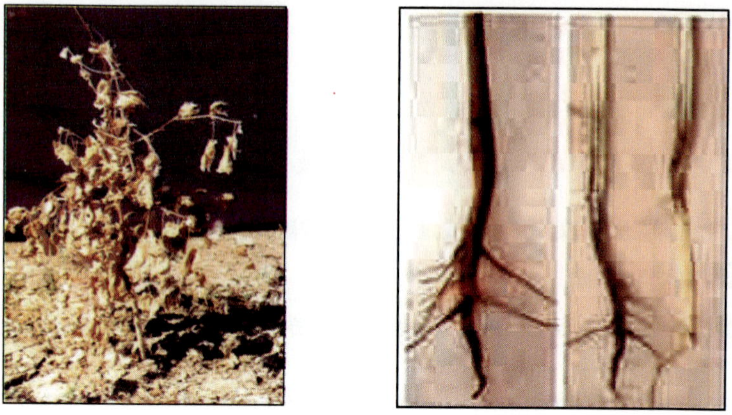

Figure 3: Root Rot Affected Plant and Roots

Pathogen

Rhizoctonia spp. viz. R. solani and *R. bataticola*

Management

1. Resistant Varieties

Grow resistant varieties eg *G. hirsutum* lines KH-33-146, 15-KW-2 (MB) and 9-kw-2 (MB) have been found to be resistant to both *R. solani* and *R. bataticola.*

2. Agronomic Practices

Mixed cropping is recommended for checking the disease intensity. Intercropping and adjusting the time of sowing of the crop are helpful in lowering the soil temperature and disease incidence. Application of FYM and cellulose containing compost which increases microbial activity, helps reduce disease incidence.

3. Biological Control

Seed treatment with *Pseudomonas fluorescens* increased seedling survival. Pellet seeds with spores of *T. viride* and *T. harzianum.*

4. Seed Treatment

Treatment of seeds with fungicides like Thiram, Bavistin, Carbendazim, Quintogene or Oxathiin is effective in reducing both seed and soil- borne infection.

5. Soil Drenching

Pre and post-sowing drenching of soil with Carbendazim and Quintogene provided maximum control. Apply PCNB containing preparations like Brassicol to the soil.

Leaf Curl Virus of Cotton

P.K. Chakrabarty, R.L. Chavhan and Rita Dhanole

Central Institute for Cotton Research, PB No. 2, Shankarnagar, Nagpur – 440 010

Symptoms

The initiation of disease is characterized by small vein-thickening on young upper leaves of plants. Upward or downward curling of leaves and development of laminar out growth of various sizes and shapes over vein on the abaxial side of the leaves are other characteristic symptoms. In severe cases reduction of inter-nodal length occurs resulting in stunting and reduced flowering and boll development. The virus also causes extensive reduction in seed yield and deteriorates quality of

Figure 1: Leaf Curl

Figure 2: Vein-thickening

cotton. A decline in the productivity of cotton observed in north zone since 1994-95 has chiefly been attributed to CLCuV disease. The disease has also been observed in the state of Karnataka on *G. barbadense* cotton grown in kitchen gardens in 1996. Reports of its occurrence in Tamil Nadu have also come recently. However CLCuV strains reported from South India appear to be restricted in nature, non-aggressive and are not spreading unlike North Indian strains. The virus spreads through its whitefly vector. In addition to cotton the virus also infects several alternate and collateral hosts, which serve as reservoirs for inoculum and provides continuity to its life cycle season after season.

Figure 3: Enation

Pathogen

Cotton leaf curl virus (CLCuV).

Management

1. Resistant Varieties

Grow resistant varieties of cotton (*G. hirsutum*) *viz.*, F1861, H1117, RS2013 and those belonging to *G. arboreum* and *G. herbaceum e.g.* LD-27, RG8, HD1, CNH123, CNH1012.

2. Agronomic Practices

Remove weed hosts of the virus from the field.

3. Chemical Control

Control whitefly by spraying appropriate insecticides.

4. Integrated Management

A study of inheritance of resistance, resistance breeding and transgenic approaches will pave way for the development of resistant cultivars.

Fig

Fig Rust

B.S. Solunke

Principal, Vivekanand Mahavidyalaya, Aurangabad, Maharashtra

Symptoms

Fig rust is an important fungus disease that attacks the leaves of figs. Pathogen infects on lower side of leaves and produces whitish – light brown coloured small spots contain millions of uredospores. In rainy season infection becomes epidemic. Increased intensity of pathogen defoliates and plant becomes naked. Fruits are also infects, shows small size, black stripes and stony texture have no market value.

Pathogen

Cerotelium fici

Management

1. Take only one Bahar (Meetha Bahar *i.e.* October – April) in a year.
2. Prune the orchard lightly in a year helps in minimization of diseased branches. This helps to rise new branching where no infection was found.
3. Remove the diseased leaves, branches, fruits, pruned material, dried stalks, weeds helps to minimize the inoculum of pathogen.
4. Spray 1 per cent Bordeaux mixture after pruning and after 10 – 15 days when new foliage starts, spray the carbendazim @ 100 gm/100-lit of water.
5. After above two sprayings, apply third spray of carbendazim rothalonil or mancozeb @ 300 gm/100-lit water which helps in complete management of rust.

Ginger

Rhizome Rot of Ginger

Shalini D. Sagar and Srikant Kulkarni

Department of Plant Pathology, University of Agricultural Sciences, Dharwad – 580 005

Symptoms

1. Soft rot symptoms observed were foliar yellowing was clearly seen starting from the margins of the lower most leaf and later progressed upwards. Later, the infection spreads to the rhizome. The affected rotten tissues emitted foul smell. Such plants came off with a gentle pull. In a matured plant, foliar yellowing and stunted growth of the plants were observed and rhizomes were rotten.
2. Wet rot is caused by *Fusarium solani*. The infected plants exhibited foliar yellowing. The affected plants wilted and dried up but didn't fall on the ground in contrast to soft rot and bacterial wilt affected ginger. The affected rhizomes showed soft and watery symptoms.

Figure 1: Water Soaked Lesions

Figure 2: Sclerotium Rot

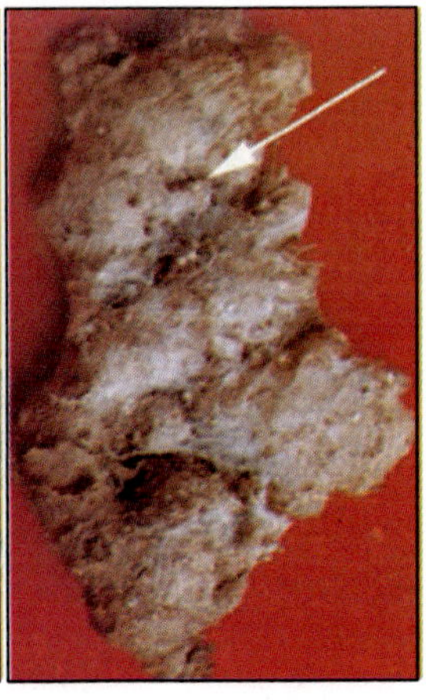

Figure 3: Root Knot

3. Bacterial wilt is caused by *Ralstonia solanacearum*. Water soaked patches or linear streaks were observed on the collar region of the pseudostem. Later leaves became flaccid with intense yellowish bronze colour was noticed and such plants drooped down. The leaves showed roll up symptoms and the whole plant dried up. Pseudostem came off easily with a gentle pull. Such rhizomes when pressed bacterial exudates ooze came out.
4. Sclerotium rot is caused by *Sclerotium rolfsii*. When rhizomes were examined there was, white mycelial growth and formation of sclerotial bodies were observed.
5. Root knot is caused by *Meloidogyne arenaria*. The infected plants in the field showed stunted and chlorotic symptoms. Small galls were seen on the roots of infected plants. The galls resembled bead-like structures.

Pathogen

Sclerotium rolfsii, Ralstonia solanacearum and *Meloidogyne arenaria*.

Management

1. Maran and Vardha ginger varieties are moderately resistant ones. CO-1 and Krishna varieties of turmeric 0020 are moderately resistant ones.
2. Early sowing reduces disease incidence.
3. Crop rotation in ginger with paddy helps reduction of disease incidence.
4. Soil solarization + Rhizome treatment with Metalaxyl MZ @ 0.3 per cent + Soil application of *T. harzianum* @ 10 kg along with 25 t FYM/ha + Soil application of *Eupatorium* @ 10t/ha

Soft Rot/Rhizome Rot of Ginger

Y.R. Sarma

Indian Institute of Spices Research, Calicut

Symptoms

The disease starts with water soak lesions on the pseudostems at the ground level. Infection gradually spreads upwards and downwards into the rhizome. Fresh sprouts are more vulnerable for the infection. Root infection is also noticed. The affected clumps exhibit foliar yellowing starting from the base and gradually spreading to the upper regions. Disease spreads through soil water or through infected rhizomes planted in the field. This results in total loss of rhizomes.

Soft Rot of Ginger

Pathogen

Pythium aphanidermatum and *Pythium myriotylum*.

Management

1. Use of Resistant Varieties

All the cultivated varieties have been found to be susceptible and no source of resistance has been identified. However, varieties like Maran, Nadia have been reported tolerant.

2. Alteration of Crop Practices

The pathogen is seed and soil borne but predominantly seed borne. Selection of healthy seed rhizomes from totally disease free gardens is the major approach to avoid seed borne infection. Mancozeb treated seed rhizomes are stored in pits lined with sand in sheds is recommended to avoid seed borne infections. Immediately after planting the beds are covered with leaf mulch to ensure better sprouting. Selection of plumpy, healthy seed rhizomes is important. Raising the crop in raised beds and providing better drainage are important to reduce the chance of infection. Citing the infected clumps early and removal of these from the main beds is utmost important to check the spread of the disease.

3. Biological Control

Trichoderma harzianum was found highly antagonistic to *Pythium*. Inoculum raised in broken sorghum meal/coffee pulp/tea waste/talcum is used for seed dressing. At the time of planting after seed-dressing 50gm of inoculum (10^6 and above cfu) is mixed with 1 kg of neem cake and applied per each bed (3 x 1m) is distributed into the pits of 4 rows with 10 pits each.

4. Chemical Control

Selected seed rhizomes are given a dip in Mancozeb solution (2gm/litre/0.2 per cent) for about 30 minutes and shade dried. Metalaxyl Mancozeb (Ridomil Mancozeb 72WP – 1.25gm/litre of water) is also used in place of Mancozeb as a seed treatment. Whenever disease is sited the infected clumps are removed from the beds and all the surrounding clumps are selectively drenched with Mancozeb or Ridomil Mancozeb or Copper oxy chloride (2gm/litre/0.2 per cent)

5. Integrated Management

Soil solarization of the beds prior to planting for about 30 days (using polythene tarping on moistened beds) has been found to be effective in reducing the disease incidence. There was a synergetic reduction in the disease when soil solarization was combined with application of biocontrol agents *T. harzianum*. However it is cost effective only for seed rhizome production. Selection of healthy seed coupled with soil application of organics like neem cake/pongamia cakes/ground nut cake @ 2 tones per hectare was effective when combined with biocontrol agents like *T. harzianum* @ 50gm per 3 x 1m bed.

Grapes

Downy Mildew of Grapes

T.S. Thind

Department of Plant Pathology, Punjab Agricultural University Ludhiana, Punjab

Symptoms

Greenish yellow oily spots appear on upper surface of leaves, which on the lower side are covered with whitish downy fungal growth. Under favourable weather conditions, uniform whitish growth is seen on the under surface of leaves. Leaves with many active spots dry up and drop pre-maturely and may adversely affect the new growth in the next season. Infected inflorescence may turn oily yellow to brown and infection on young peduncles results in the death of entire cluster. Infected clusters may develop downy growth under high humidity.

Pathogen

Plasmopara viticola

Management

1. There should be restriction on the movement of planting material at regional, national and international level since the pathogen spreads through dormant buds.
2. Infected plant material after pruning must be removed and burnt before bud sprouting to reduce primary inoculum.
3. Spacing of vines, training system and row direction should be carefully planned so as to ensure maximum air circulation and minimum leaf wetness duration.
4. Give protective application of fungicides such as Bordeaux mixture, copper oxychloride, cupric hydroxide, chlorothalonil etc. and repeat at 7-10 days interval. Repeat application after rains.
5. Under heavy disease risk situations due to favourable weather conditions, give sprays of systemic fungicide mixtures such as metalaxyl+mancozeb, dimethomorph+mancozeb, curzate+mancozeb and fosetyl-Al etc. at 10-12 days interval.

Anthracnose of Grapes

T.S. Thind

Department of Plant Pathology, Punjab Agricultural University Ludhiana, Punjab

Symptoms

Symptoms appear on all the green parts of the vine including fruits. Small, irregular reddish brown to dark brown spots appear on the leaves, which turn grey in the centre. Sometimes, the necrotic tissue drops out of the lesion leaving shot holes. Young leaves are more susceptible. On twigs, tendrils and petioles, brown sunken spots appear which later mature into cankers with dark purple and raised margins. During moist weather, pink fungal spore masses appear within the lesions. The affected shoots show restricted growth. On the berries, the spots are initially light brown but later turn grey in the centre and are surrounded by reddish and purple margins, thus looking like 'birds eye' spots. The effected berries may crack.

Pathogen

Gloeosporium ampelophagum (Elsinoe ampelina)

Management

1. All canes showing cankers should be removed at pruning time and burnt. This will help in reducing primary inoculum during the growing season. Ground level canes and branches should also be removed. Give one spray of Bordeaux mixture immediately after pruning.
2. Spray vineyards at the time of leaf emergence with thiophanate methyl (0.1 per cent), carbendazim (0.1 per cent) or Bordeaux mixture. At least four sprays of fungicides should be given during rainy season at 10-14 days intervals. Care should be taken not to spray the same fungicide repeatedly.
3. Variety Delight is tolerant whereas Bharat Early and Hussaini are resistant.

Powdery Mildew of Grapes

T.S. Thind

Department of Plant Pathology, Punjab Agricultural University Ludhiana, Punjab

Symptoms

Whitish powdery patches appear on upper as well as lower leaf surface. Powdery patches are also produced on stem, tendril and young fruit branches leaving dark discoloration on affected areas. More prominent symptoms develop on young fruits as white powdery growth. Affected berries turn dark in colour, may develop cracks and do not develop further.

Pathogen

Uncinula necator

Management

1. Use training system that reduces dense growth and allows proper air circulation through the canopy. This will not only reduce the disease severity but also allows better penetration of fungicide sprays.
2. Fungicides like wettable sulphur, dinocap, triadimefon, myclobutanil, flusilazole and penconazole are very effective in checking powdery mildew infections. Their use should begin during early stages of vine development after bud burst. Spray schedules of 7-10 days interval are usually required for obtaining effective control by sulphur while dinocap can be applied at 10-14 days interval. EBI fungicides are commonly used at 2-3 week interval.
3. Orchard sanitation is important in reducing disease pressure during growing season.
4. Avoid excessive use of nitrogenous fertilizers as it promotes succulent growth and increases disease severity.

Anthracnose of Grapes

A.S. Deokate

U.B. Group of Companies, Baramati, Dist. Pune, India

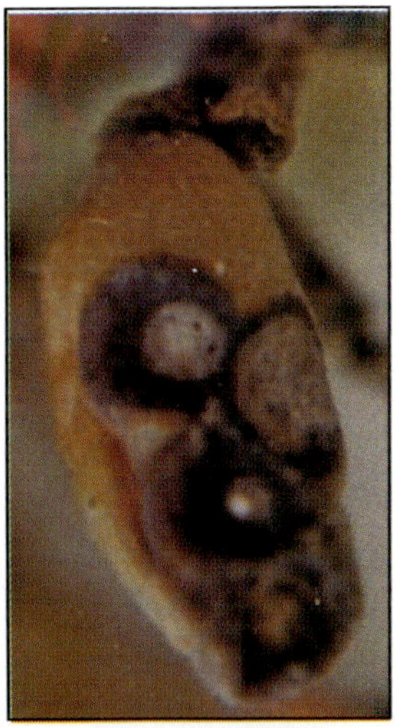

Symptoms

1. Infected leaves show necrotic symptoms.
2. On berries, black rough spots are produced at the earlier stages and spoil them.

Pathogen

Gloeosporium ampelophagum (Pass.) Sacc.

Management

1. Application of carbendazim (0.02 per cent) and Score alone or in mixture with copper oxychloride or Bordeaux mixture, alternate use of these gives better control.

2. Extracts of plants like *Emblica officinalis* and *Terminalia chebula* alone or in mixture with carbendazim (2 ug/ml).
3. Biological control with *Trichothecium roseum* or *Trichoderma* sp.
4. Keep the orchard clean by uprooting the weeds.

Groundnut

Dry Root Rot of Groundnut

Avinash B. Ade

Dr. Babasaheb Ambedkar Marathwada University, Aurangabad

Symptoms

Pre emergence and rotting of seeds, rotting of the hypocotyls, post emergence seedling blight.

Pathogen

Aspergillus niger

Management

1. Use of resistance varieties: Resistant variety: J-11

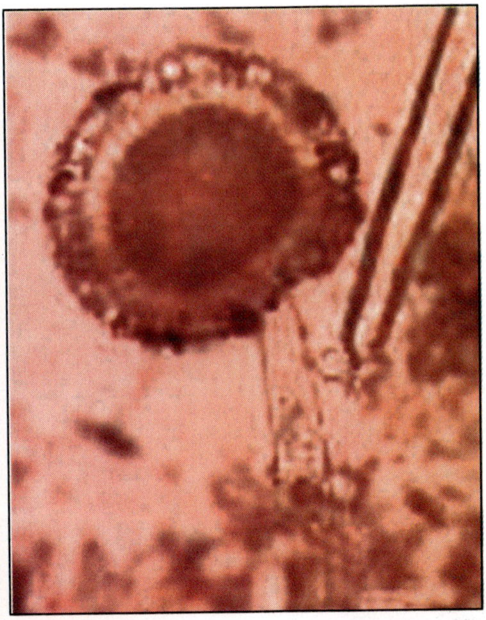

2. Alteration of crops: Rotation of groundnut with chickpea and wheat.
3. Chemical control:
   (*a*) Seed treatment with tetramethyl thiram disulphide (thiram TMTD) 75 per cent WP @ 3.5g/Kg seed.
   (*b*) Captan 80 per cent WP 3 g/Kg seed or carbendazim (Bavistin 50 per cent WP @ 2g/Kg seed, control seed borne infection.
   (*c*) Spray mancozeb 3 g/lit of water immediately when disease is noticed.
4. Biocontrol: Use of *Pseudomonas aeruginosa, Pseudomonas fluorescens* and *Rhizobium*.

Aflaroot of Groundnut

Avinash B. Ade

Dr. Babasaheb Ambedkar Marathwada University, Aurangabad

Symptoms

1. Seed and un-emerged seedlings attacked by the fungus are rapidly reduced to a shrivelled, dried or brown or black mass covered by yellow or greenish spores.
2. After seedlings emergence cotyledons already infected with the fungus show necrotic lesions with reddish brown margins. Under field conditions the diseased plants are stunted and often chlorotic.

Pathogen

Aspergillus flavus

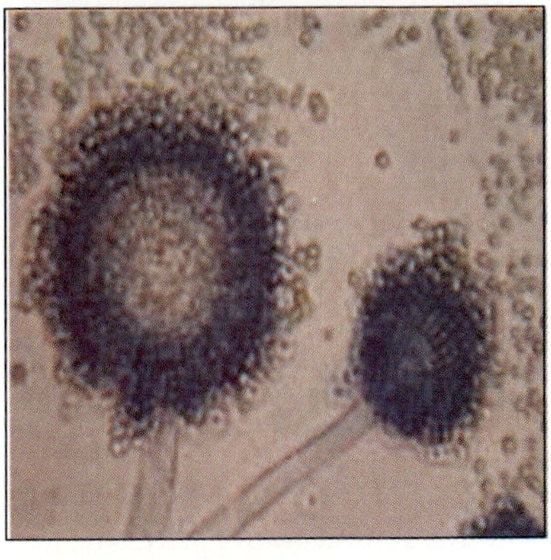

Management

1. Chemical Control

Seed treatment with carbendazim @ 2 gm/Kg seed or Captan @ 3 gm/Kg seed or thiram @ 3-5 gm/Kg seed has given good results.

2. Agronomic Management

Since the fungus is a weak parasites, agronomic practices which favour rapid germination and vigorous growth of seedling will reduce the chance of *Aspergillus flavus* infections.

3. Biocontrol

Application of *Pseudomonas fluorescens* and *Rhizobium*

Rust of Groundnut

B.M. Kareppa

Department of Biotechnology, D.S.M. College, Parbhani – 431401

Symptoms

Pathogen

Puccinia arachidis

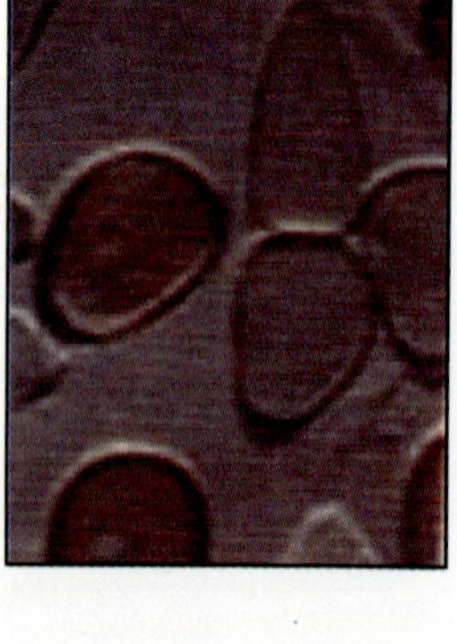

Management

Use of resistant varieties, fungicides, biocontrol agents and integrated management.

Tikka/Leaf Spot of Groundnut

B.M. Kareppa

Department of Biotechnology, D.S.M. College, Parbhani – 431401

Symptoms

Pathogen

Cercospora arachidis

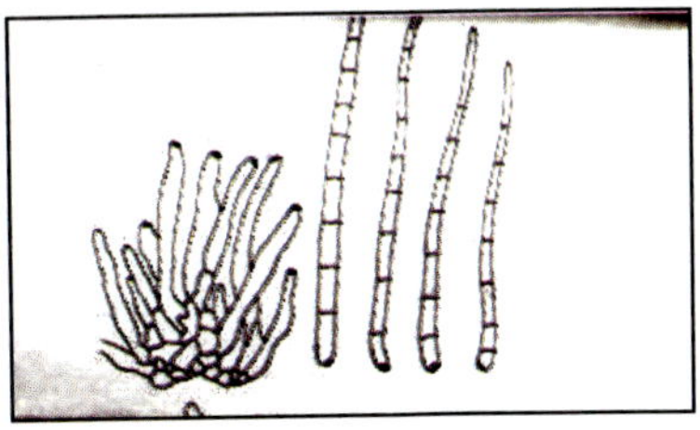

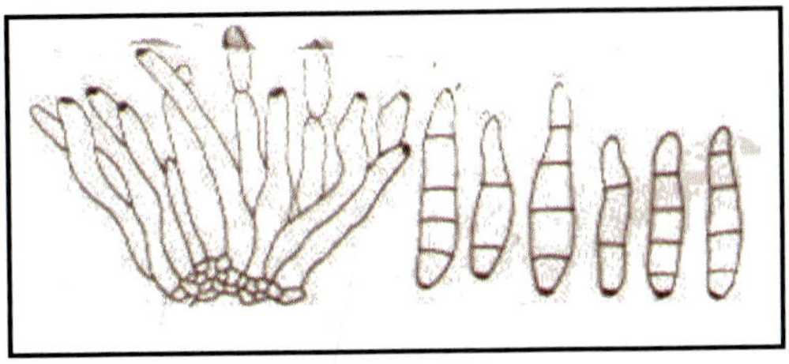

Management

Clean cultivation, use of resistant varieties, fungicides, biocontrol agents and integrated management.

Alternaria Blight of Groundnut

D.C. Kamthane[1] and B.M. Kareppa[2]

[1]SGBS College, Purna Dist. Parbhani, M.S.
[2]Department of Biotechnology, D.S.M. College, Parbhani – 431401

Symptoms

Pathogen

Alternaria sp.

Management

Use of resistant varieties, fungicides, biocontrol agents and plant leaves extract.

Guava

Wilt Disease of Guava

D.K. Chakrabarti

Department of Horticulture, NDUAT, Faizabad

Symptoms

The symptoms vary with variation of edaphic factors. In alkaline soil with moderate soil moisture wilting takes place slowly (slow wilting) (Figure 1) and affected plants take about one year to die completely. Prior to wilting the leaves become chlorotic followed by defoliation and drying of young twigs. But in sandy loam slightly acidic soil of riverside orchards the wilting takes place abruptly (sudden wilting) (Figure 2) within 15 days of the first appearance of flaccidity on some tender twigs. The leaves shrivel and die on the tree that gives a scorched appearance. Sometimes slow wilting symptoms appear only on one side of the grown up tree (partial wilting). Tree barks split and peel off. Xylem vessels of roots are blackened and clogged with the mycelium.

Figure1: Slow Wilting Symptoms on Guava

Pathogen

A number of soil borne fungi have been reported to be associated with the wilt affected plants. But in India, three prominent fungi *viz. Fusarium oxysporum* (Sheld.) Snyd and Hans f.sp. *psidii, F. solani* (Mart.) Sacc., and *Macrohomina phaseolina* (Tassi) Goid have been consistently isolated and their pathogenicity has been confirmed. Experimental evidence demonstrated that they act synergistically as a complex.

Management

1. Resistant Varieties

Lucknow 49 (Sardar guava), Dholak Sind, Nasik, Supreme. Some *Psidium* species like *P. friedrichs-thalianum* (Chinese guava), *P. molle, P. cujavilis* are resistant and may be used as root stocks.

2. Alternation of Crop Practices

The plants may be raised over Chinese guava rootstocks. In October garlic crops may be grown at the basin of the tree.

Figure 2: Sudden Wilting Symptoms on Guava

3. Biological Control

Application of *Trichoderma viride* or *Aspergillu niger* (250g; cfu 2 10^7/g) at the rhizosphere soil immediately after rainy season (August-September).

4. Chemical Control

For recovery of the infected plants, soil drenching with 10-15 liters solution of carbendazim (0.2 per cent) per tree in March, June and September has been recommended.

5. Integrated Management

The integrated management strategy includes application of 2 kg each of gypsum and margosa (neem) cake along with 1 kg lime per tree of 8 years of age or above before onset of rainy season. The other components of the IPM are use of bio-agents, companion crops (garlic) with allelopathic effects, soil drenching with carbendazim and use of resistant rootstocks as described above.

Wilt of Guava

H.K. Kehri, Varun Khare and Sudha Singh

Department of Botany, University of Allahabad, Allahabad – 211 002

Symptoms

The symptoms of the disease include browning and wilting of the leaves, discoloration of stem and death of the branches along one side. The susceptibility of the disease decreases with the age. Infected parts develop chlorosis followed by wilting of entire seedling and leaf abscission. Sometimes, infection girdles the entire stem and the whole plant may wilt. The leaves may fall off and the whole plant may

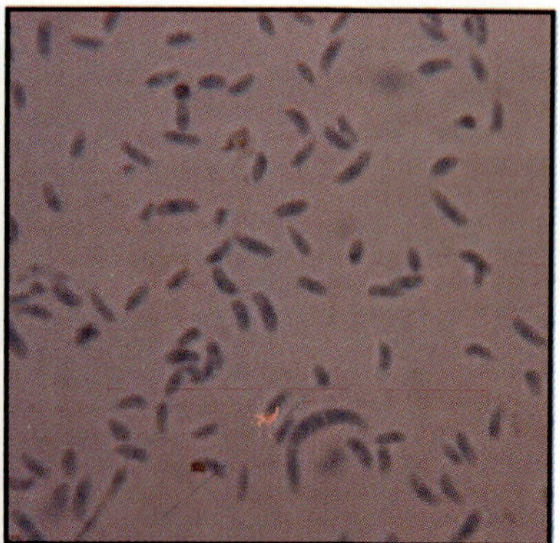

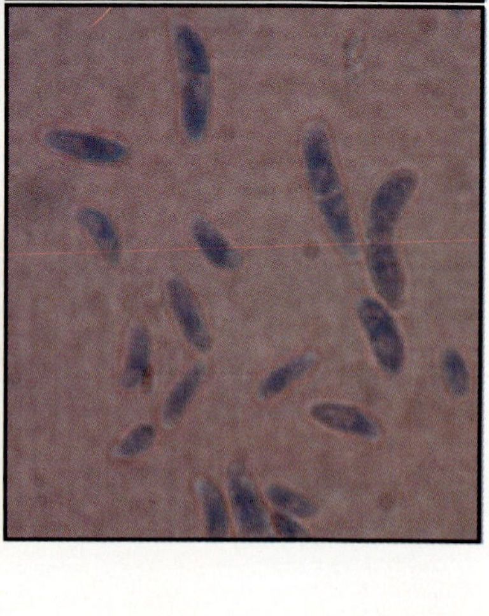

die with in a year. Presence of hyphae in xylem vessels of roots of the wilted seedlings is also reported.

Pathogen

Fusarium oxysporum f. sp. *psidii.*

Management

There is effective measure known for this soil-borne disease. Diseased twigs are cut and burnt. Wilted plants are uprooted and the empty pits are disinfected by fire using foliage. Soil treatment with lime or gypsum @ 1.3 kg/tree. For this treatment, roots must first be exposed to a depth of 15 cm. infection of apparently healthy plants with 0.1 per cent 8-quinolinol sulphate may be provided protection against wilt for at least one year. Balanced nutrition of host reduces severity of infection especially when the source of nitrogen is organic. VAM fungi (*Glomus aggregatum*) acts as a potential biocontrol agent against wilt of guava. Use of resistant variety is also desirable *viz.* Banarasi, Dholka Sindh, Nasik, White Guava No. 6299, Clone 32-12, Supreme, Webber, Hart Rolf and Lucknow-49.

Anthracnose of Guava

H.K. Kehri, Varun Khare and Sudhir Chandra

Department of Botany, University of Allahabad, Allahabad – 211 002

Symptoms

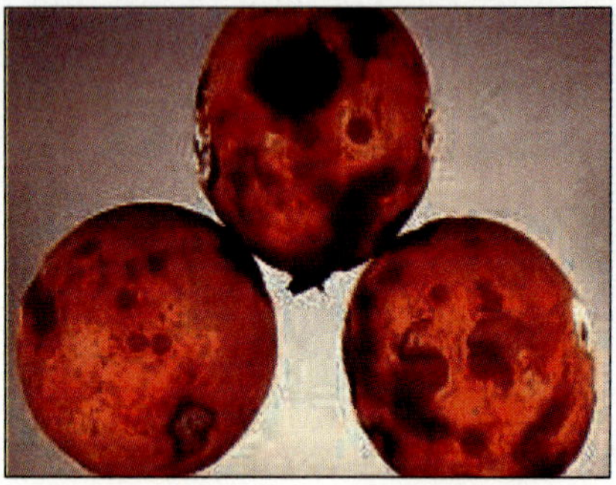

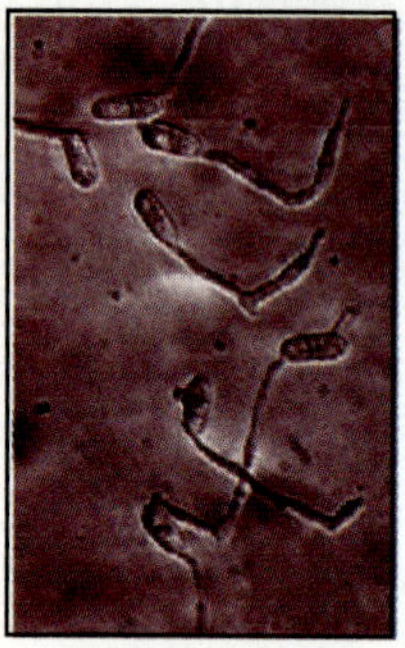

The disease mainly affects all the aerial parts especially young shoots and leaves leading to the shedding of the young buds, flowers and fruits. Growing tips turn dark brown, followed by black necrotic areas that extend backwards causing typical symptoms of dieback. From twigs, pathogen attacks petioles and young leaves, which become distorted with dead areas at margin. Ultimately the dried twigs are left without any leaves. Small pin head sized spots appear on the unripe fruits, which gradually enlarge to form brown or black sunken spots presenting the characteristic symptom of the disease. Infected areas on fruits become hard and corky and develop black stromata in the center of the lesions. Affected fruits develop malformations and shed prematurely.

Pathogen

Gleosporium psidii Delacr.

Management

The disease incidence can be minimized by spraying the tree with the Bordeaux mixture (3:3:50). Besides this, copper oxychloride, difoltan and dithane Z-78 are also effective against disease. Irradiation of fruits with gamma rays (100 kr) after 5, 18 and 24 hours of inoculation has been found suitable for control of the rot. Sodium metabisulphate and stable bleaching powder is also effective.

Fruit Canker of Guava

H.K. Kehri and Varun Khare

Department of Botany, University of Allahabad, Allahabad - 211 002

Symptoms

Disease symptoms are most evident on leaves and fruits. Affected leaves show red or brown spots, which are either angular or irregular in shape. Fruits spots are dark brown, rough, corky and raised. Affected fruits remain small sized and develop malformations and cracks. Canker is not known to penetrate deeply into the fruits, but infection leads to the rapid loss of ascorbic acid content from the fruit.

Pathogen

Pestalotia psidii Pat.

Management

Spraying Bordeaux mixture or Lime Sulphur can check disease. Variety Safeda and Apple Guava are highly resistant against the canker disease. It can also be checked by post harvest wash with Aureofungin or dip treatment with Aretan. Two homeopathic drugs, *viz. Kali iodatum* 87 and *Arsenicum album* 181 (5 minute dip) were effective both as protectant as well as therapeutant in controlling rot. A wash with potassium permanganate (1 per cent) or boric acid (2 per cent) could protect the fruits.

Stylar End Rot or Phomopsis Rot of Guava

H.K. Kehri and Varun Khare

Department of Botany, University of Allahabad, Allahabad – 211 002

Symptoms

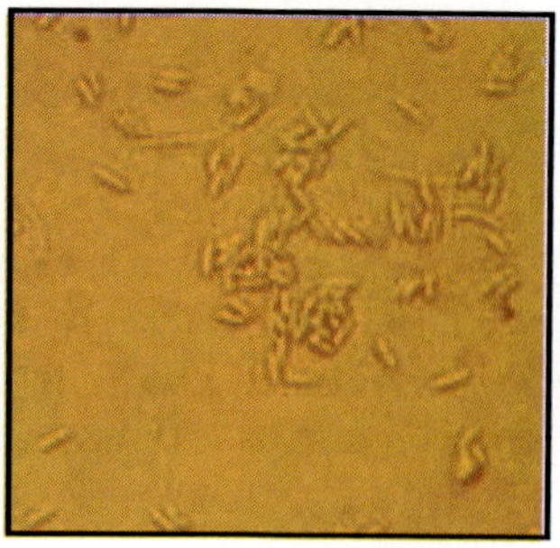

The disease appears at the stylar-end of the fruit as water soaked lesions. These lesions enlarge and often coalesce to form larger lesions. The affected portion gets darker in color. With the advancement of the disease, small white or large brown pycnidia appear on the lesions.

Pathogen

Phomopsis psidii De Camara.

Management

Amongst fungicides, Ziram (0.25-1.0 per cent) has been found to be effective against phomopsis rot. Benlate, a systemic fungicide, at 2000-3000 ppm alone or with 0.2 per cent paraffinic mineral oil and Tecto-40, a formulation of Thiabendazole (systemic fungicide) at 2000 ppm was effective in saving the fruits from rots. Arya (1988) has reported that Bavistin, Difolatan, Calyxin (1000 ppm) and Dithane M-45 (1250 ppm) are highly effective in checking the disease. Mustard oil and leaf extracts of *Eucalyptus* sp. and *Aegle marmelos* (50 per cent conc.) are also effective in checking the rots.

Botryodiplodia Rot of Guava

H.K. Kehri and Varun Khare

Department of Botany, University of Allahabad, Allahabad – 211 002

Symptoms

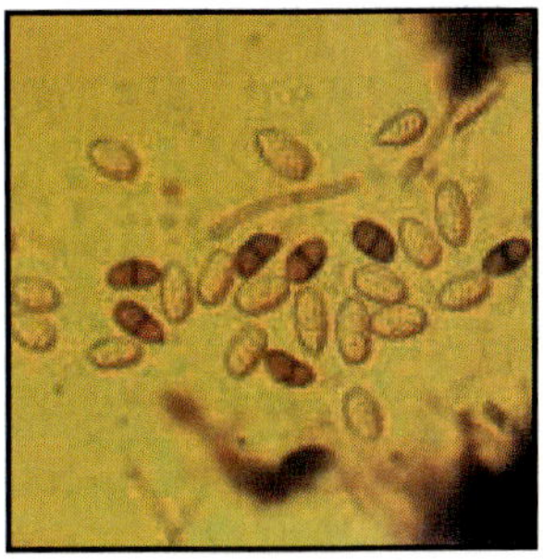

Botryodiplodia rot usually initiates from the stalk end of the fruits as light brown spots. These spots enlarge and proceed downwards with irregular margin. Often major portion of the fruit gets rotted and becomes soft. Sometimes, ooze (watery sap) comes out from the lesions. Later on small dark colored pycnidia develop on the lesions.

Pathogen

Lasiodiplodia theobromae (Pat.) Griff. and Maubl. (*Botryodiplodia theobromae* Pat.).

Management

Fruits stored at 10°C escaped rotting for ten days but stored fruits when transferred to room temperature, however, developed rot. Storing fruits at 15°C can also minimize loss due to this rot. Careful handling and quick disposal of fruits can minimize the losses.

Wilt of Guava

A.K. Misra

Central Institute for Subtropical Horticulture, Lucknow

Symptoms

The symptom may be either quick wilt or slow wilt. Quick wilt is considered as the typical wilt in guava. The affected plants show leaf yellowing with slight curling at the terminal branches, later becoming reddish and subsequently shedding of leaves take place. Twigs become bare and fail to bring forth new leaves or flowers and eventually dry up. Fruits of all the affected branches remain underdeveloped, become hard, black and stony. The entire plant becomes defoliated and eventually dies. Few plants also show partial wilting, which is a very common symptom of wilt in guava.

Pathogen

The exact cause of the disease is still not fully understood. Various pathogens reported by different workers are *Fusarium oxysporum* f. sp. *psidii, F. solani, Macrophomina phaeseoli, Rhizoctonia bataticola, Cephlosporium* sp., *Gliociadium roseum* and *Verticillium albo-atrum* etc. However, wilt due to *Fusarium oxysporum* (Fr.) Schl. f. sp. *psidii* Prasad, Mehta and Lai is more accepted.

Management

1. Use of Resistant Varieties

The cultivars Apple Colour, Chittidar, Seedless and Spear Acid, Superior Sour Lucidium, Red Flesh and Smooth Green as tolerant. F1 population of *Psidium molle* X *Psidium guajava* was also identified free from wilt and recommended as resistant root stock.

2. Alteration of Crop Practices

Field sanitation, uprooting of wilted plants, trench around tree trunk of affected plants, thrash burning in the pit, pit treatment with formalin, balanced nutrition, separate channel irrigation to individual plant, green manuring, protection of root damage during tillage, intercropping with turmeric or marigold, soil solarization with 30fm transparent polyethylene sheet during May-June are some of the methods to check wilt. Various workers suggested following chemical controls.

(*a*) Soil treatment with carbendazim or thiophanate methyl

(*b*) Soil treatment with 1.82 kg lime or Gypsum/tree

(*c*) Judicious amendments of N and Zn.

(*d*) Application of 6 kg Neem cake + 2 kg Gypsum/plant.

These chemicals are suggested but no chemical control has been found effective once the plants are affected.

3. Disease Management through Bio-control Agents

Application of enriched FYM with *Aspergillus niger* strain AN 17 @ 5kg/pit at the time of initial planting and 10 kg per basin in grownup plants is effective in controlling wilt of guava. Other bioagents *viz. Trichoderma viride, Trichoderma harzianum* and *Penicillium citrinum* have also been found effective.

4. Integrated Eco-friendly Approach

Considering the complexity of the problem, following integrated eco-friendly approach for the control of guava wilt was suggested. Utilizing bioagent *Aspergillus niger* strain AN 17, using resistant root stock (*P. molle x P. guajava),* intercropping of guava with turmeric or Marigold as well proper field sanitation to minimize losses due to the disease.

Mango

Powdery Mildew of Mango

A.K. Misra

Central Institute for Subtropical Horticulture, Lucknow

Symptoms

Powdery mildew of mango is a widespread and most important disease of mango. It attacks inflorescence, stalk of the inflorescence, leaves and young fruits. The characteristic symptom of the disease is the white superficial powdery growth of the fungus on these parts having millions of conidia, which are borne in chains on conidiophores. The affected flowers fail to open and may fall prematurely. Young fruits are covered entirely by the white mildew growth, resulting dropping of fruits. When it affects the slightly more grown fruits, epidermis of the infected fruits cracks and corky tissues are formed. Young leaves are attacked on both the sides as small irregular grayish patches and infection is restricted to the area of the central and lateral veins of infected leaves and such leaves often twist, curl and get distorted.

Pathogen

Oidium mangiferae Berthet.

Management

1. Use of Resistant Varieties

Latif, Hurr. Neelum, Zardalu, Banagalora, Totapari, Khurd, Janardan Pasand, Ghanya, Jahangir, Yellamondela Thiamandi, Bablipunasa, Kharbuja, K.O. 7, Oliveira, Imperial and Carlota were reported resistant.

2. Alteration of Crop Practices

Pruning of diseased leaves and malformed panicles (on which powdery mildew survives during off season) reduce primary inoculum load and improve the control achieved by spraying of mildewcides.

3. Chemical Control

As the inflorescence infection causes actual serious harm, 3 sprays of fungicides during a flowering season is recommended at 15-20 days interval.

(*a*) First spray of wettable sulphur (0.2 per cent) when panicles are 3-4" in size.

(*b*) Second spray of dinocap (0.1 per cent) after 15-20 days of first spray.

(*c*) Third spray of tridemorph (0.1 per cent) after 15-20 days of second spray.

4. Integrated Management

Removal of infected foliar infection during January and chemical control during flowering period successfully control the disease.

Anthracnose of Mango

A.K. Misra

Central Institute for Subtropical Horticulture, Lucknow

Symptoms

The disease cause leaf spot, leaf blight, wither tip, blossom blight and fruit rots.

On Leaves

Characteristic symptoms appear as oval or irregular vinaceous brown to deep brown spots of various sizes scattered all over the leaf surface, later forming elongated brown necrotic areas measuring 20-25 mm in diameter. The lesions get blighted and rupture and show 'shot hole' symptom. Young leaves are more prone to attack.

On Twigs

The tip of the very young branches, start drying from tip downwards.

On Blossom

Small black spots appear on the panicles and open flowers, which gradually enlarge and coalesce to cause death and drop of flowers.

On Fruits

On stored fruits, black round spots are produced which later coalescence to form large irregular botches or even cover the entire fruit. The spots have large deep cracks. Under moist conditions, the blackened areas are covered with minute pinkish reproductive bodies of the fungus. Staining, russetting and tear streaking, involving only the skin of the fruit, are attributed due to the same fungus.

Pathogen

Colletotrichum gloeosporioides Penz. [*Glomerella cingulata* (Stons.) Spauld and Schrenk]

Management

1. Use of Resistant Varieties

Cultivars Edward, Mayaguazano and Elamandi are reported resistant against anthracnose.

2. Alteration of Crop Practices

Diseased leaves, twigs and fruits, lying on the floor of the orchard, should be collected and all infected twigs from the tree should be pruned and burnt.

3. Chemical Control

Blossom infection by anthracnose can be controlled effectively by 2 sprays of carbendazim (0.1 per cent) at 15 days interval, while the foliar infection can be controlled by spray of 0.3 per cent copper oxychloride. For the control of post harvest fruit infection, pre-harvest sprays of 0.1 per cent thiophanate methyl or carbendazim followed by post harvest hot water treatment at 52 ±1°C for 30 min. have been found effective. The duration of hot water treatment can be reduced to 15 min by adding carbendazim (0.05 per cent).

4. Integrated Management

The management strategies recommended to control anthracnose include cultural practices, tree management, variety selection and protective fungicide spraying.

Malformation of Mango

A.K. Misra

Central Institute for Subtropical Horticulture, Lucknow

Symptoms

Two distinct types of symptoms are described. These are vegetative malformation and floral malformation.

Vegetative Malformation

The characteristic symptoms in seedlings are loss of apical dominance and swelling of vegetative buds in the leaf axil or at the tip with small shoot-lets bearing small scaly leaves, giving bunchy top appearance. These buds persist even when dried and form girdle around the branches due to crowding of nodes.

Floral Malformation

The symptoms of floral malformation are characterized by deformation of panicles, suppression of apical dominance, shortening of primary and secondary axis, giving the panicle a characteristic clustered appearance, thickened rachis of panicles, and greatly enlarged flowers with large disc and the preponderance of staminate flowers bearing scanty pollen grains. The rachis of malformed panicle is more branched, thicker and heavier than the normal ones. Such panicles continue to grow till the advent of rains, afterwards dry up and turn in to a black mass, which may persist on the tree up to the next flowering season.

Pathogen

Fusarium moniliforme var. *subglutinans.*

Management

1. Use of Resistant Variety

Bhadauran, a monoembryonic variety and cultivar Ellaichi were found resistant and as source of resistance to mango malformation.

2. Alteration of Crop Practices

Deblossoming of early emerging panicles either alone or in combination with NAA (100-200 ppm) at the time of flower bud differentiation in October reduce floral malformation. Elevation of temperature around the mango trees by covering trees with colourless polythene tent raise the av. temperature by 3°C around the tree as well as internal temperature of leaves by 1°C resulting in early emergence and faster growth of panicles and inhibition of malformation.

3. Chemical Control

Spray of carbendazim at 15 days interval during October to February reduces the incidence of malformation. Spray with NAA (100-200 ppm) at the time of flower bud differentiation in October reduces the incidence of malformation.

4. Integrated Disease Management

A strategy of integrated disease management of mango malformation may be as follows:

(*a*) Developing resistant variety by using resistant source (cv. Ellaichi or Bhadauran).

(*b*) Application of NAA during the month of October.

(*c*) Application of carbendazim 0.1 per cent at 15 days interval during October to February.

(*d*) Raising temperature of orchard with orchard heaters during January month.

(*e*) Deblossoming of early-emerged flowers.

(*f*) Regular removal of malformed panicles.

Malformation of Mango

D.K. Chakrabarti

Department of Horticulture, NDUAT, Faizabad

Symptoms

The disease has three distinct types of symptoms *viz.* bunchy top (Figure 1), vegetative (Figure 2) and floral (Figure 3) malformation. In bunchy top, one or several bunches are formed at the tip or lower down the main shoot of young plants. The bunch is formed of numerous thickened, small shootlets and leaflets arising from a leaf axil and clustered together. Most of the diseased plants die prematurely. Vegetative malformation induces in young plants excessive vegetative branches thickened with short internodes, restricted growth and rudimentary leaves. In malformed panicles flowers are far more in number, compactly arranged, enlarged, greener and mostly staminate and remain unopened. Pollens show poor viability and malformed panicles keep on growing even after the fruit season. Depending upon weather conditions malformed panicles may be less compact. But they hardly set any fruit.

Figure 1: Floral

Figure 2: Bunchy Top

Figure 3: Vegetative malformation symptoms on mango

Pathogen

Fusarium subglutinans (Wollenweber and Reinking) Nelson, Toussoun and Marasas (=*F. moniliforme* Sheldon var. *subglutinans* Wollenweber and Reinking in section Liseola).

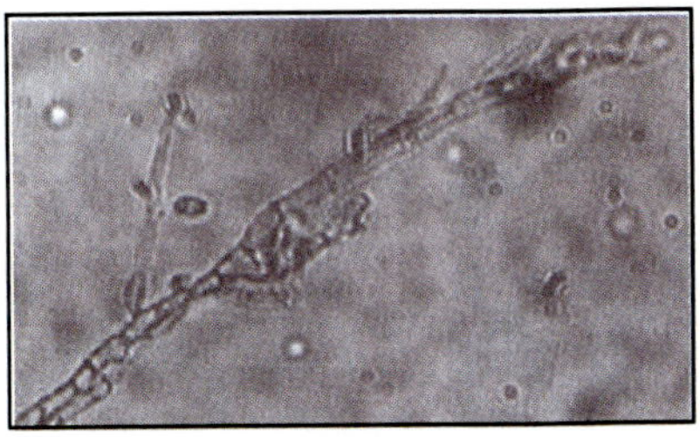

Figure 4

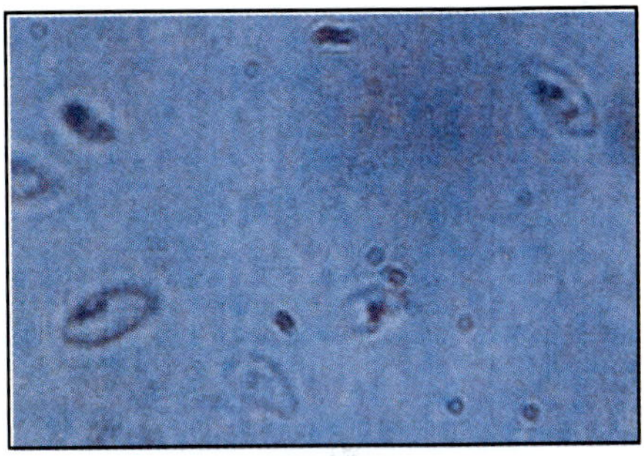

Figure 5

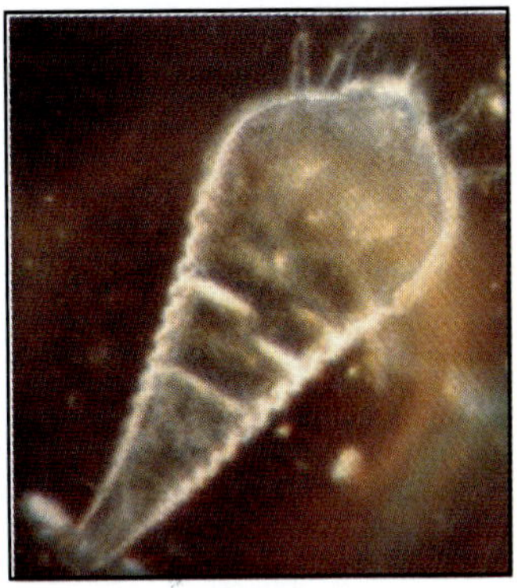

Figure 6

Figure 4: Polyphialides; Figure 5: Microconidia of *Fusarium moniliforme* var. *subglutinans*; Figure 6: Eriophyes *(Aceria mangifera)* Mite, the Vector

Management

1. Resistant Varieties

The cultivars Bahadauran and Elaichi show field resistance but have less market demand.

2. Alteration of Crop Practices

A mixed cultivation of different cultivars specially with regular and alternate bearers, instead of mono blocks, both the disease incidence and its spread in the orchard become less.

3. Biological Control

In nature *Aspergillus niger* van Tiegh acts as antagonist by parasitizing the pathogen which was also confirmed in some preliminary trials

4. Chemical Control

Sprayings of 200 ppm napthyl acetic acid (Planofix, 40ml in 9 liter of water) during first week of October followed by deblossoming at bud burst stage or 4 sprayings of copper oxychloride (0.25 per cent) during vegetative period or 3 sprayings of captan (0.1 per cent) before flowering up to fruit set stage.

5. Integrated Management

The integrated management strategy recommends removing malformed plant parts one feet below from the point of attachment twice – once after harvesting of fruits (before onset of rainy season) and next in October before flower bud initiation. The emerging buds of autumn and spring flushes should have prophylactic sprayings with copper fungicide (0.2 per cent) and acaracide (dimithoate 0.05 per cent).To replenish the hormonal and micronutrient deficiencies, sprayings of napthyl acetic acid (200 ppm) before flower bud initiation (October-November) and chelated zinc sulphate (40 ppm) at developing stage of panicles (January-February) have been recommended. For raising healthy nurseries, scion shoots are sprayed with Captan (0.2 per cent) thrice *i.e.* immediately after defoliation of scion shoots, 24h and 7 days after grafting. The grafted plants then are shifted to poly house (temperature 25-31°C, RH >85 per cent) for further growth. At the non-availability of poly house, grafting may be done during spring flush and nurseries should be maintained far away from infected orchards under adequate sunlight.

Powdery Mildew of Mango

P.A. Fugro and K.P. Vaidya

Department of Plant Pathology, College of Agriculture,
Dr. Balasaheb Sawant Konkan Krishi Vidyapeeth,
Dapoli – 415 712, Dist. Ratnagiri, M.S.

Symptoms

Disease generally appears as whitish or grayish powdery areas on tender leaves and inflorescence. Powdery mass consists of fungal conidia and conidiophores. Infection spreads from tip of inflorescence and covers floral axis, young leaves and stem. Disease attacks flowers before fertilization and fruits in their earliest stages. The affected fruits drop off prematurely causing heavy losses in fruit yield.

Pathogen

Oidum mangiferae Berth.

Management

A number of sprays with different fungicides for the control of the disease have been recommended. However, recent studies with two potent fungicides viz; Amistar 25 EC and Score 25 EC have brought about more encouraging results. The application of any of these fungicides @ 0.1 per cent concentration (3-4 sprays at an interval of 10-12 days on initiation of the disease) has resulted in significantly lower disease incidence and higher fruit yield. However, in addition to these sprays one prophylactic spray with any of these fungicides just at the initiation of third blossom flush in Konkan has been proved to be more effective.

Some of the mango varieties like Neelam, Benglora, Ratna, etc. have shown some tolerance to powdery mildew.

Mungbean & Urdbean

Mungbean and Urdbean Yellow Mosaic Virus

O.D. Kohire

Marathwada Agricultural University, ARS, Badnapur, Dist. Jalna, M.S.

Symptoms

Appearance of yellow specks or spots on young leaves, with the next emerging trifoliate leaf exhibiting irregular yellow and green patches and slight purkening and reduced leaf size. The yellow patches coalesce to form larger patches that develop into yellow mottle; the entire leaf may turn yellow. In addition to the yellow mottle this a necrotic centre develops in the yellow spots in some varieties. Yellow mottle symptoms are very clear on the pods, while the rest of the plant may not show any symptoms.

Pathogen

The disease is caused by Mungbean yellow mosaic virus (MYMV) and a whitefly *Bemisia tabaci* acts as a vector.

Management

1. Use of Resistant Varieties

Mungbean – ML 131, ML 267, PM 1, PM 2, PM 3, PM 4, PDM 54 (Moti), PDM 139 (Samrat), BM 4, Mum 2, Narendra Mung 1, TARM 1, Pusa Bold 1 and Ganga 8, Urdbean – PDU 1, IPU 94-1, Pant U 19, UG 218, To 94-2, LBG 17, WBG 26, KU 300.

2. Alteration of Crop Practice

Adjustment of sowing dates to avoid the disease. Inter/mixed cropping of Mungbean and Urdbean with non host crops like sorghum, pearl millet and maize.

3. Biological Control

A fungus *Paccilomyces farinosus* has been found parasitic on the vector *Bemisia tabaci.*

4. Chemical Control

Application of Phorate, Disulfoton granules @ 1-2 kg a.i./ha in furrows during sowing. To control white flies spray of Metasystox, Malathion (0.1 per cent) Dimethoate (0.02 per cent), at the initiation of the disease.

5. Integrated Management

IDM components like use of resistant varieties, cultural practices and judicious use of chemicals may reduce vector population and disease incidence.

Cercospora Leaf Spot of Mungbean and Urdbean

O.D. Kohire

Marathwada Agricultural University, ARS, Badnapur, Dist. Jalna, M.S.

Symptoms

Water soaked lesions appear on the leaf blade and soon the affected tissues turn brown to reddish brown. Well defined spots often bound by veins and purplish border develop, the centres of which may turn gray, powdery growth due to the sporulation of the fungus is seen at the centre of the spots. Though the spotting is mostly confined to leaf blades. In some cases, the affected tissue is killed and the dead portion drops out leaving shot hole symptoms on the leaves.

Pathogen

Cercospora cruenta (*Mycosphaerella cruenta*), *Cercospora canescens*

Management

1. Alteration of Crop Practice

Field sanitation, crop rotation, destruction of infected plant debries and avoiding collateral hosts in the vicinity of the crop. Adjustment of date of sowing recommended to control the disease and mulching reduce primary inoculum of the pathogen thereby decreasing disease incidence.

2. Chemical Control

Two spray of Bavistin (300 g/ha) at 30 and 45 days after sowing significantly controlled the disease.

3. Integrated Management

Use of resistant varieties, and seed from disease free crop. Adjustment of sowing dates. Destruction of affected plants as soon as noticed in the field.

Powdery Mildew of Mungbean and Urdbean

O.D. Kohire

Marathwada Agricultural University, ARS, Badnapur, Dist. Jalna, M.S.

Symptoms

A white powdery growth occurs on the leaves, spreading to cover the stem and other plant parts. The infection takes place at any stage of plant growth, but more severely when the plants are flowering and it persists until harvest. In very severe cases the entire plant becomes affected and wilts.

Pathogen

Erysiphe polygoni DC

Management

1. Use of Resistant Varieties

Mungbean – TARM 1, TARM 2, WGG 46, WGG 47, ML 3, ML 131, SS 1, AVRDC 1381, LM 294. **Urdbean** – Warangal No. 1792, LBG 17, No. 6203, No. 65-1 and S 905-1-1.

2. Alteration of Crop Practice

Delayed sowing of Mungbean and Urdbean with wider spacing has been found to considerably reduce the disease severity. Adjusting the date of sowing with wider spacing can reduce powdery mildew incidence.

3. Biological Control

Hyperparasitism of fungus *Ampelmyces quisqualisces* on powdery mildew fungus in urdbean.

4. Chemical Control

Several fungicide such as Karathane, Calaxin, Bavistin, Benlate, Topsin M,

Bayleton, Sulphur dust and Inodofil Z-78 have been found effective to control the disease under field condition.

5. Integrated Management

IDM components like resistant varieties, cultural practice and judicious use of chemicals may reduce disease incidence.

Okra (Bhindi)

Yellow Vein Mosaic of Okra (Bhindi)

V.C. Khilare

Vasantrao Naik Mahavidyalaya, Aurangabad – 431 003

Symptoms

The disease is characterized by a homogenous interwoven network of yellow veins enclosing islands of green tissue within. In extreme cases, the infected leaves become totally yellow or cream colour. Infected plants remain stunted and bear very few deformed and small fruits. The disease causes heavy loss in yield if the plants get infected within 20 days after germination.

Pathogen

Yellow vein mosaic virus (YVMV).

Management

1. Among the pesticides/bio-pesticides applied, imidacloprid significantly reduced the whitefly population. Neem extract and biological control agent (EM) were also effective against *Bemisia tabaci.*
2. The vector population and disease incidence can be reduced by application of 4 to 5 foliar sprays of either Dimethoate (0.05 per cent) or Metasystox (0.02 per cent) or Monocrotophos (0.05 per cent), followed by 1 or 2 sprays of mineral oil (2 per cent).
3. One soil application of Furadon @ 1.5 kg a.i./ha at the time of sowing seeds is also recommended.
4. Infected plants must be removed from the field
5. Some okra lines have been evolved at IIHR, Bangalore, which are completely resistant to yellow vein mosaic of okra.

Onion

Purple Blotch of Onion

A.H. Rajasab

Plant Pathology Laboratory, Gulbarga University, Gulbarga – 585106

Symptoms

Lesions appear first as small whitish sunken spots on leaves or scapes, which later turn brown. As the lesions enlarge, they become purplish with clear zonations. The margins are reddish surrounded by yellow zone that extends both upwards and downwards for some distance. The lesions then quickly girdle the leaves/scape making them to droop. *A. porri* is also reported to cause umbel withering. Bulbs are

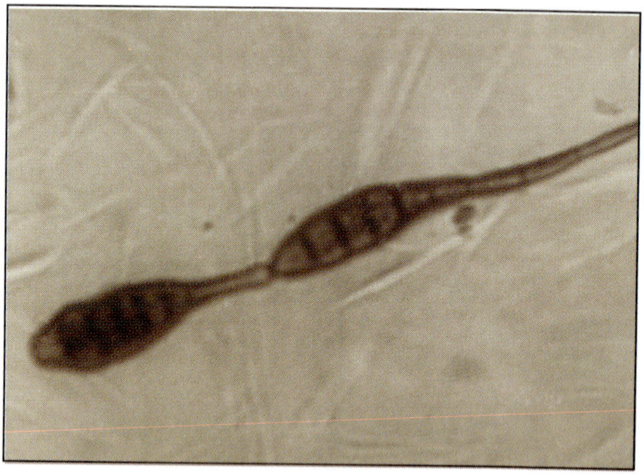

Figure 1: *A. porri* Conidia

Figure 2: Symptoms on Leaves

also affected, they turn gradually from yellow to tan red and decay finally. Affected bulb scales become papery.

Pathogen

Alternaria porri

Management

Applying of four sprays of Mancozeb @ 0.3 per cent with Monocrotophos @ 0.05 per cent controlled purple blotch and thrips populations on leaves resulting in improved yield in the variety N-53. Other recommended formulations of chemicals used for disease control are:

1. Application of four sprays of Mancozeb @0.25 per cent at fortnightly intervals during kharif season.
2. Spraying of Mancozeb (0.2 per cent) and copper oxychloride (0.3 per cent) thrice at fortnightly intervals.
3. Spraying of Mancozeb at 0.25 per cent at 15 day intervals along with sticker Triton.
4. Five sprays of Dithane M-45 @ 0.25 per cent at 15 days intervals along with sticker Triton.
5. Application of Hexaconazole in seed treatment (0.2 per cent), root dip (0.1 per cent) and foliar spray (0.1 per cent).

Opium Poppy

Downy Mildew of Opium Poppy

D.K. Chakrabarti

Department of Horticulture, NDUAT, Faizabad

Symptoms

The disease manifests in two forms: i) leaf spot and ii) severely infected plants (systemic infection). In leaf spot (Figure 1), initially slightly bleached or yellow spots appear on leaf blades; develop in all directions in various shapes and later on turn dark brown to black due to necrosis. The spots coalesce giving a burning appearance. Infected portions become brittle; lower leaves dry up and soon fall. But plants do not die nor leaves are malformed. In the severely infected plants (Figure 2), the leaves remain small, chlorotic, curling downward and closely packed. Heavy sporulation occurs on entire lower surface (Figure 3) and also on peduncle, buds and capsules. Buds remain under developed and do not open.

Figure 1: Blighted Leaf

Figure 2: Systemically Infected Plant

Figure 3: Sporulation on Lower Surface of Leaf

Pathogen

Peronospora arborescens (Berk.) de Bary

Figure 4

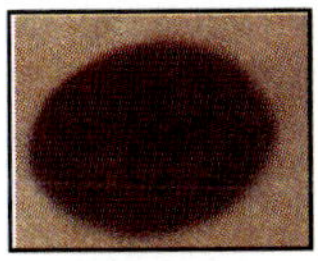

Figure 5

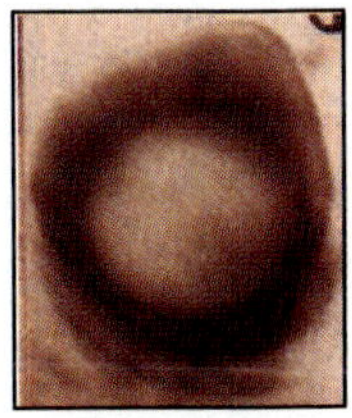

Figure 6

Figure 4: Conidiophore; Figure 5: Conidia ; Figure 6: Chlamydospore of *P. arborescens*

Management

1. Resistant Varieties

A cultivar Gazipur Local (Figure7) shows high resistance and being used in breeding programme.

Figure 7: Resistant cv Gazipur Local

2. Alteration Crop Practices

Measures like crop rotation for three years, removal of infected plants and debris, destruction of a common weed collateral host, *Argemon mexicana* (Figure 8), sowing of seeds within first week of November, proper balance between nitrogenous fertilizer (90-12 kg N_2O/ ha) and FYM (100-200 q/ha) and avoidance of excess soil moisture significantly reduce the disease incidence.

Figure 8: Colateral Host, *Argemon mexicana*

3. Biological Control

Seed treatment with biofertilizer, Azotobactor reduces requirement of nitrogen fertilizer as well as disease incidence. Seed treatment with fungal antagonist, *Trichoderma viride* (10g/ Kg seeds, 2 10^7 cfu/g) followed by drenching of 15-day old seedlings with water suspension of the antagonist (100/litre) provide effective control.

4. Chemical Control

Seed treatment with metalaxyl (Apron SD 10g/kg) and its subsequent foliar sprayings (Ridomil MZ, 0.25 per cent) at 35, 55 and 75 days after of sowing effectively control the disease. Dithane M-45 is also effective against the pathogen.

5. Integrated Management

An integrated management strategy consists of improved cultural practices as mentioned above, seed treatment and drenching of 15-day old seedlings with *T.viride* and sprayings of metalaxyl at 35 and 55 days after sowing provides maximum disease control.

Pigeonpea

Phytophthora Blight of Pigeonpea

Sumia Fatima

Aurangabad College for Women, Aurangabad – 431 001, M.S.

Symptoms

Symptoms start 3 weeks after sowing near water-logged soil. Blight occurs near the soil and spread towards the upper part of the plant.

Pathogen

Phytophthora drechsleri Tucker f.sp. *cajani*

Management

1. Use of resistant variety Hy – 3c.
2. Seed treatment with metalaxyl.
3. Use of Ridomil 25 WP as foliar spray.
4. Treat the soil with Ridomil 1G, 2G or 5G granules.

Wilt of Pigeonpea

O.D. Kohire

Marathwada Agricultural University, ARS, Badnapur, Dist. Jalna, M.S.

Symptoms

The typical Symptoms are drying of plants and wilting. The leaves show mild chlorosis and yellowing before wilting. Wilted plants produce light brown or purple streaks rising from the base upwards on the main stem. Browning or blackening of the wood is seen, if the bark from the streak area is pealed off.

Pathogen

Fusarium udum

Management

1. Use of Resistant Varieties

BSMR 736, BSMR 853, Maruti (ICP 8863), Asha (ICPL 87119), MA 3, Narendra Arhar, MAL 13.

2. Alteration of Crop Practice

Inter/mixed cropping and crop rotation with non host crops have great potential in management of the disease. Intercropping of pigeonpea with sorghum has been reported as most effective. The best ratio of Sorghum: Pigeonpea (1:1).

3. Biological Control

Management of wilt through antagonist *viz., Bacillus subtilis, Trichoderma harzianum, Trichoderma viride* is an attractive and ecofriendly approach.

4. Chemical Control

Carbendazim + Thiram (1:2 g/kg seed) Benlate, Benomyl + Thiram have been found very effective.

5. Integrated Management

Removal of studies, crop rotation for 3 years and other cereals timely sowing and keep spacing normal, intercropping with Sorghum (1:1) and host plant resistance with MAL 13, Narendra, Arihant and Amar.

Sterility Mosaic of Pigeonpea

O.D. Kohire

Marathwada Agricultural University, ARS, Badnapur, Dist. Jalna, M.S.

Symptoms

Stunting of plants, reduction of size and mosaic mottling of leaves, profuse branching and pale green foliage colour. In the field, the diseased plants can be seen in patches which are bushy and pale green, severely diseased plants do not produce flowers and pods and turn are completely sterile. In case of late infection, partial cessation of reproductive structures occurs.

Pathogen

Tenui virus (pigeonpea sterility mosaic virus). The causal agent is transmitted by a mite; *Aceria cajani*. The disease spreads through infective mites disseminated by wind.

Management

1. Use Resistant Variety

BSMR 853, BSMR 736, Sharad, Pusa 9, Narendra Arhar 1, Amar, MA 3, MA 6, ICPL 87051.

2. Alteration of Crop Practice

Destruction of off-season volunteers plants from the vicinity. Discouraging perennial or rationed crop may reduce the off-season survival of vector and pathogen inoculum in disease escape during the subsequent main crop season.

3. Biological Control

No useful information is available on different component of management of pigeonpea sterility mosaic.

4. Chemical Control

Seed dressing with 10 per cent aldicarb @ 3 g/kg has been found to protect the crop upto 45 days after sowing. Spraying of acaricides like Kethane, Metasystox or Diamethoate at 0.1 per cent.

5. Host Plant Resistance

Cultivation of resistant varieties is the best way for management of SMD.

6. Integrated Management

Removal of ratooned or perennial pigeonpea in the vicinity, use resistant variety, seed treatment, crop rotation 3 years with cereals, sowing and spacing normal with foliar spray of acaricide.

Bacterial Blight (Oily Spot) of Pomegranate

V.T. Jadhav, K.K. Sharma and Jyotsana Sharma

National Research Centre on Pomegranate, Solapur, Maharashtra

Symptoms

Figure 1: Nodal/Twig Blight

Pathogen

Xanthomonas axonopodis **pv. punicae** (Hingorani and Singh) Vauterin, Hoste, Kersters and Swings.

Figure 2: Bud and Flower Blight

Figure 3: Leaf Spots

Management

1. None of the commercial grown cultivars is resistant.

Figure 4: Fruit Spots with Cracking

2. Use disease free planting material.
3. Follow clean cultivation and orchard sanitation.
4. Avoid rainy season crop.
5. Spray Streptocycline (250-500ppm), Bordeaux mixture (0.5 per cent), Copper oxychloride (0.25 per cent) effectively check disease if proper spray schedules are followed.

Wilt of Pomegranate

K.K. Sharma, Jyotsana Sharma and V.T. Jadhav

National Research Centre on Pomegranate, Solapur, Maharashtra

Symptoms

Figure 1: Partially Wilted Plant

Figure 2: Dark Discolouration of Wood

Pathogen

Wilt pathogen *Ceratocystis fimbriata* producing two types of conidia (Figure 3) and perithecia with long necks (Figure 4)

Fusarium sp., Phytophthora sp and *Rhizoctonia* are also found associated in some areas. Shot hole borer and nematodes also are seen associated in some areas. These may help to spread the pathogen.

Management

1. Follow plant spacing of 4.5 m x 3.0 m
2. Plant pomegranate in sandy loam soil with proper drainage

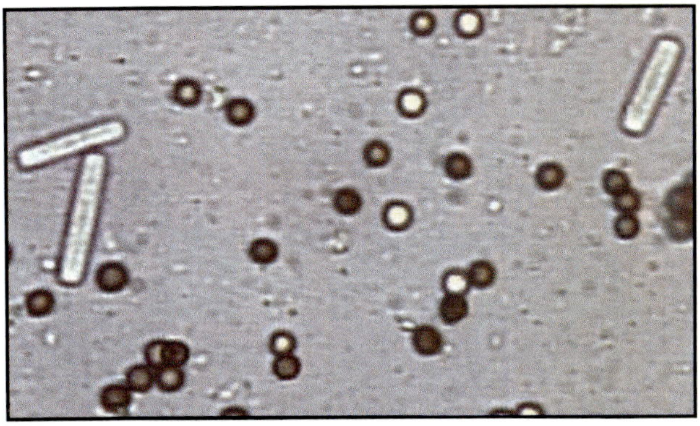

Figure 3

Figure 4

3. Drench soil with carbendazim (0.2 per cent) or propiconazole (0.15 per cent) or Tridemorph (0.15 per cent) + Chlorpyriphos (0. 25 per cent) before planting in diseases prone areas is advisable
4. The completely wilted plants should be uprooted and burnt and drench the plant basin as above.

Cercospora Leaf and Fruit Spots of Pomegranate

Jyotsana Sharma, K.K. Sharma and V.T. Jadhav

National Research Centre on Pomegranate, Solapur, Maharashtra

Symptoms

Figure 1: Reddish Brown Spots on Leaves

Pathogen

Conidia of pathogen *Cercospora punicae* (Figure 3).

Figure 2: Black Spots on Fruits

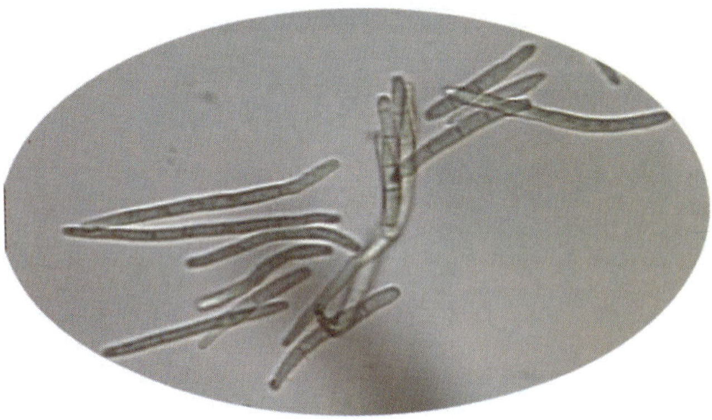

Figure 3: Conidia of Pathogen *Cercospora punicae*

Management

At flower initiation or disease appearance, spray the crop with any one of the fungicides carbendazim (0.1 per cent), mancozeb (0.25 per cent), or copper oxychloride (0.25 per cent).

Anthracnose of Pomegranate

Jyotsana Sharma, K.K. Sharma and V.T. Jadhav

National Research Centre on Pomegranate, Solapur, Maharashtra

Symptoms

Figure 1: Fruit Spots with Reddish Purple Border and Light Brown Centers

Pathogen

Conidia of pathogen *Colletotrichum gloeosporioides* (Figure 2).

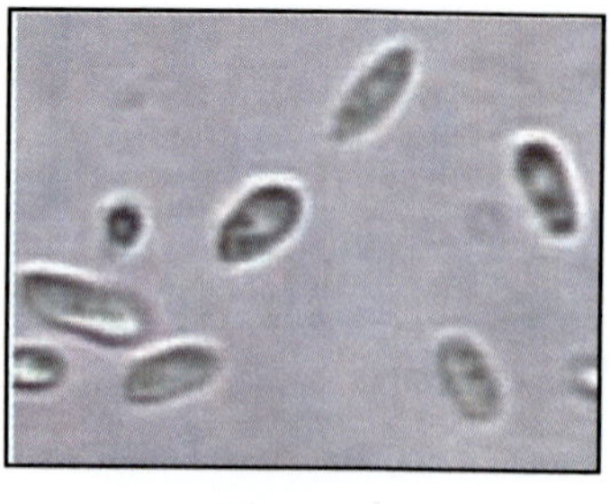

Figure 2

Management

At flower initiation or disease appearance, spray the crop with any one of the fungicides carbendazim (0.1 per cent), mancozeb (0.25 per cent) or Copper oxychloride (0.25 per cent).

Sun Scald of Pomegranate

V.T. Jadhav, K.K. Sharma and Jyotsana Sharma

National Research Centre on Pomegranate, Solapur, Maharashtra

Symptoms

Figure 1.

Factors Responsible

High temperature along with excessive light, drought, and low relative humidity are usually responsible for sun scald injuries.

Management

1. Damage is more in open canopies so avoid very heavy pruning and develop proper canopy.
2. Spray processed Kaolin (Inert clay) thrice at 15 days interval during hot summer months. First spray is of 5 per cent and next two of 2.5 per cent. If heavy rain or winds occurs spray interval can be accordingly reduced.

Figure 1: Brown/Bronze Colour Skin Due to Sun Scald

Abiotic Fruit Cracking of Pomegranate

V.T. Jadhav, K.K. Sharma and Jyotsana Sharma

National Research Centre on Pomegranate, Solapur, Maharashtra

Symptoms

Figure 1.

Figure 1: Arils Exposed Due to Splitting of Fruits

Factors Responsible

Improper irrigation and boron deficiency lead to cracking. Calcium and Potash deficiency may also cause cracking.

Management

1. Irrigate the plants with adequate quantity of water at regular intervals
2. Before harvest spray Boron @ 0.2 per cent (2g/L)
3. Apply calcium and potash as per soil test values

Bacterial Wilt and Brown Rot of Potato

S.M. Paul Khurana*

Vice-Chancellor, Rani Durgavati Vishwavidyalaya, Jabalpur – 482001, M.P.

Pathogen

Pseudomonas solanacearum Smith

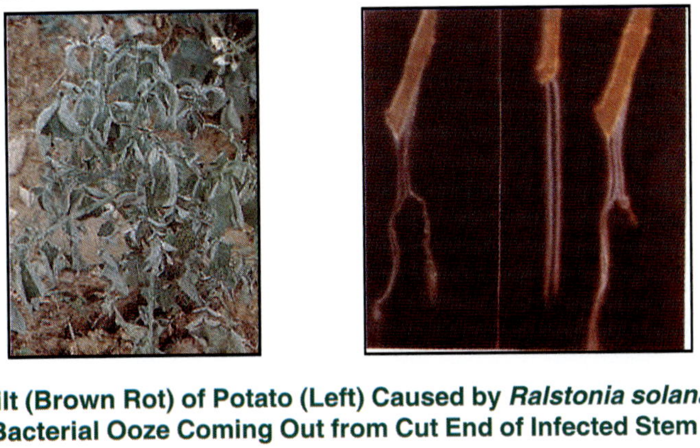

Bacterial Wilt (Brown Rot) of Potato (Left) Caused by *Ralstonia solanacearum*; Note the Bacterial Ooze Coming Out from Cut End of Infected Stem (Right)

Management

1. Wilt free seed production may be undertaken only in the disease free areas.
2. Use only healthy (disease-free) seeds. It is safer to obtain seed from known disease-free areas.
3. Do not cut seed tubers. Cutting spreads the pathogen even to healthy tubers.

---

* Former Director, CPRI, Shimla - 171001 (H.P.). The author acknowledges the help of CPRI colleagues and publications in preparation of this article.

4. Apply stable bleaching powder @ 12 kg/ha (mixed with fertilizer in furrows while planting). It reduces wilt incidence by 80 per cent.
5. Follow 2-3 years rotation with crops like maize, finger millet, cereals, garlic, lupin, onion, cabbage and sanai (*Crotolaria* sp.).
6. Restrict post-emergence tillage to minimum. Practice full earthing up immediately after planting. Collect and burn all diseased plant material including infected tubers after harvest. Soil should be ploughed twice at 20-30 days intervals and left exposed to hot summers in the plains and peninsular India during winters in the hills.

Black Scurf and Stem Canker of Potato

S.M. Paul Khurana

Vice-Chancellor, Rani Durgavati Vishwavidyalaya, Jabalpur – 482001, M.P.

Pathogen

Rhizoctonia solani Kuhn.

Severe Typical Black Scurf on a Potato Tuber

Management

1. Combination of tuber disinfection and improved cultural practices successfully check the incidence and severity of black scurf.
2. In the hills where pathogen survives in the soil, tuber treatment with MEMC (Organomercury) and soil application of PCNB @ 30 kg per ha are most effective.
3. In the North Indian plains where the pathogen has restricted survival, treatment of the diseased seed either with fungicides or chemicals like

MEMC (Organomercury), Thiabendazole; TBZ + 8 Hydroxyquinoline; acetic acid + ZnSO4; carbendazim, boric acid, effectively controls the disease.

4. Hot weather cultivation has been found to be highly effective in reducing the inoculum levels in the fields.
5. Growing maize or *'dhaincha'* (*Sesbania aegyptiaca*) for green manure as rotation crop also checks the inoculum and disease build up.

Common Scab of Potato

S.M. Paul Khurana

Vice-Chancellor, Rani Durgavati Vishwavidyalaya, Jabalpur – 482001, M.P.

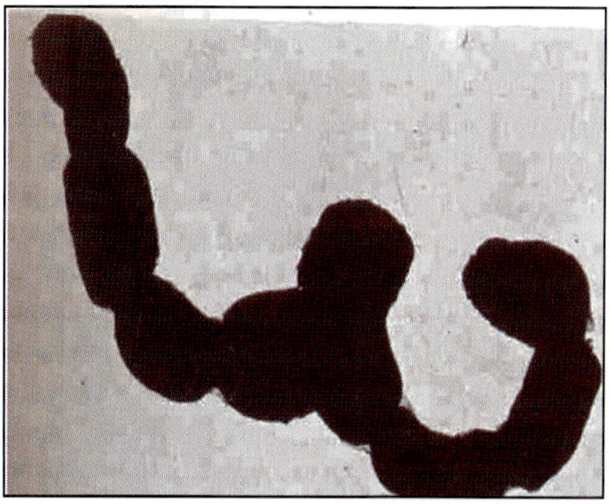

Electron Micrograph of a *Streptomyces* spp. (x 15000)

Pathogen

Streptomyces spp.

Management

The Pathogen is difficult to control because of long survival both on seed tubers and in soils. Following practices help minimize the inoculum and creating adverse conditions for pathogen spread/ disease development, hence recommended:

1. Use only disease free seed tubers for planting.

Varying Degrees of Mild to Severe Common Scab on Potato Tubers

2. Give tuber treatment with organomercurial compounds (0.015 per cent for 20 minutes) or boric acid (3 per cent for 30 minutes) before or after cold storage and dry in shade.
3. Irrigate the crop repeatedly to keep the moisture near to field capacity right from tuber initiation until the tubers measure one cm in diameter.
4. Maintain high moisture in ridges at least for a few weeks during the initial tuberization phase of crop.
5. Crop rotations with crops like wheat, pea, oats, barley, lupin, soyabean, sorghum, bajra and adopt green manuring keep the disease (pathogen) under check.
6. Plough the potato fields in April and leave the soil exposed to high temperatures during May to June in the North Indian plains.

Late Blight of Potato

S.M. Paul Khurana

Vice-Chancellor, Rani Durgavati Vishwavidyalaya, Jabalpur – 482001, M.P.

Pathogen

Phytophthora infestans.

Late Blight Affected Potato Leaf

Zoosporangiospores of *P. infestans*

Management

Plains

1. Cultivate only blight resistant cultivars like Kufri Jyoti, Kufri Badshah, Kufri Chipsona land II, Kufri Sindhuri, Kufri Jawahar, Kufri Sutlej etc should be cultivated.
2. Before storage and planting all blighted tubers must be removed.
3. Ridges should be made high enough to cover all tubers.

4. Stop irrigation as soon as the weather conditions turn congenial (temperature 10 to 20C, RH>80 per cent) for the disease appearance.
5. Give protective spray with a contact fungicide *viz.* mancozeb (0.2 per cent) before appearance of the disease and repeat at 8 to 10 days interval. Followup with use sticker like Trition AE (0.1 per cent) one or two sprays of metalaxyl (Ridomil) (0.25 per cent).

Hills

1. Only resistant cultivars should be grown.
2. Spray the crop with mancozeb (0.2 per cent) as soon as the weather becomes congenial for late blight appearance. Give subsequent sprays depending on weather and crop condition should be applied at 8 days interval till crop maturity. In severe cases, apply systemic fungicide like metalaxyl (Ridomil) (0.2 per cent) may be applied alternating with mancozeb. For sprays (two each of mancozeb and ridomil) are sufficient.
3. Rogue out Ground keepers/bolters as early as possible.
4. Cultural practices in the hills are same as outlined for plains.

Potato Leafroll

S.M. Paul Khurana

Vice-Chancellor, Rani Durgavati Vishwavidyalaya, Jabalpur – 482001, M.P.

Pathogen

Potato leafroll virus (PLRV); (syn= phloem necrosis virus).

Potato Leaf Roll (Left) Caused by PLRV and the Purified Virus Particles (Right)

Management

The disease may be managed by adopting the following measures in an integrated manner:

1. Plant only certified virus-free seed potatoes obtained from reliable sources.
2. Virus-free seed is multiplied in aphid free areas/periods coupled with practices for ensuring sanitation, rouging of infected plants, if any.
3. Buildup of incoming aphid vectors be checked/delayed to some extent by application of suitable contract/systemic insecticides.
4. Dehaulming of crop is done as soon as the vector aphid populations exceeds critical limit of 20 aphids/100 compound leaves.

Potato Mosaics

S.M. Paul Khurana

***Vice-Chancellor, Rani Durgavati Vishwavidyalaya,
Jabalpur – 482001, M.P.***

Pathogen

Majority of potato viruses (except potato leafroll and the symptomless ones) cause mild to severe mosaic, mottling/crinkle etc. as named below:

PVX:Latent or Faint Mosaics, PVA:Mild Mosaics, PVM:Paracrinkle, PVY:Severe mosaics, PAMV:Aucuba mosaics, PVX+A: Crinkle, PVX+Y:Rugose.

Severe Mosaic Caused by PVY Plus PVX (Left) and PVY Virions (Right)

Management

Production of healthy (virus-free) seed is an elaborate activity employing a package of measures comprising selection of healthy stocks and their multiplication in an area free from vector during the crop season.

The mosaic diseases can be managed by adopting the following measures in an integrated manner:

1. Planting essentially disease-free seed stocks obtained from approved or reliable sources.

2. Use of virus resistant varieties, if available.
3. In case of seed crop/plot carefully rogue out diseased plants alongwith their tubers and dispose them away from the field atleast twice during the growing season.
4. Dehaulm the seed plot/crop before the aphids cross-critical level to enforce rigid control of the insect vector.

Late Blight of Potato

R.K. Arora

Central Potato Research Station, Jalandhar, Punjab

Symptoms

Figure 1 and 2.

Figure 1

Pathogen

Phytophthora infestans (Mont.) de Bary. (Figures 3 and 4).

Figure 2

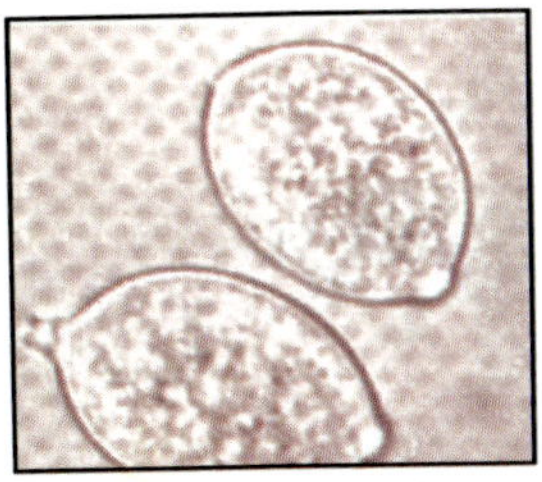

Figure 3: Sporangia

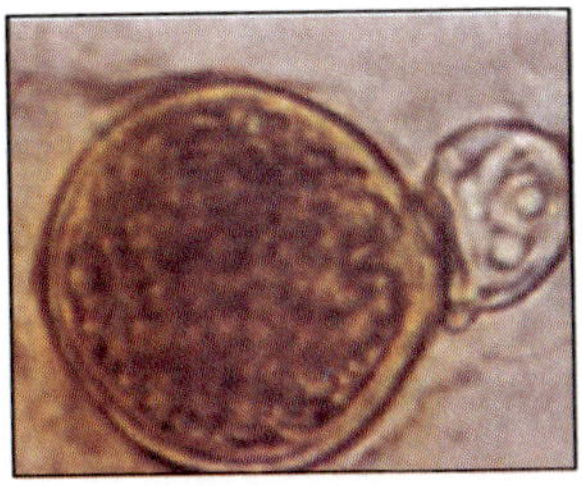

Figure 4: Oospore

Management

1. Eliminate sources of primary infection such as infected tubers, volunteer plants
2. Use disease free seed of resistant varieties such as Kufri Badshah, Kufri Megha, Kufri Giriraj, Kufri Kanchan, Kufri Swarna, Kufri Chipsona I, II and III and Kufri Phushkar
3. Provide prophylactic spray of contact fungicides such as mancozeb before closure of rows.
4. Follow a 7 to 10 day spray schedule with contact fungicides after disease appearance depending upon weather conditions. Use pre packed mixture of metalaxyl + mancozeb or cymoxanil + mancozeb at recommended dose if disease appears in severe form.
5. To protect tuber infection in soil provide thick soil cover on potato ridges, restrict irrigation after disease appearance, cut and dispose infected haulms and harvest after 15 to 20 days after haulms cutting when skin of tubers mature. Sort out and bury the infected tubers.

Early Blight of Potato

R.K. Arora

Central Potato Research Station, Jalandhar, Punjab

Symptoms

Pathogen

Alternaria solani Sorauer (Ellis and Martin)

Management

1. Poor growth predisposes the plants to leaf spots. Therefore, apply recommended dose of fertilizers to support vigorous plant growth.
2. Spray a contact fungicide such as mancozeb, captafol or chlorothalonil at recommended dose as soon the lower leaves develops the spots.

3. Avoid unnecessary sprinkler irrigation since alternate wet and dry conditions enhance build up of disease.
4. Remove diseased haulms and bury them to contain sources of inoculum.
5. Allow tubers to mature before harvest, avoid bruises and cuts at harvest and sort out the infected tubers showing brown surface discoloration.

Black Scurf of Potato

R.K. Arora

Central Potato Research Station, Jalandhar, Punjab

Symptoms

Pathogen

Rhizoctonia solani (Kuhn).

Management

1. Seed treatment with boric acid (3 per cent spray before cold storage), pencycuron, penpiclonil, *Trichoderma viride* (before planting) found effective for the control of seed inoculum.
2. Soil solarization with transparent polyethylene mulching for 4 weeks during hot summer months reduces soil borne inoculum.
3. Avoid late harvesting of the crop since late crop show higher disease incidence.

Dry Rot of Potato and Fusarial Wilt

R.K. Arora

Central Potato Research Station, Jalandhar, Punjab

Symptoms

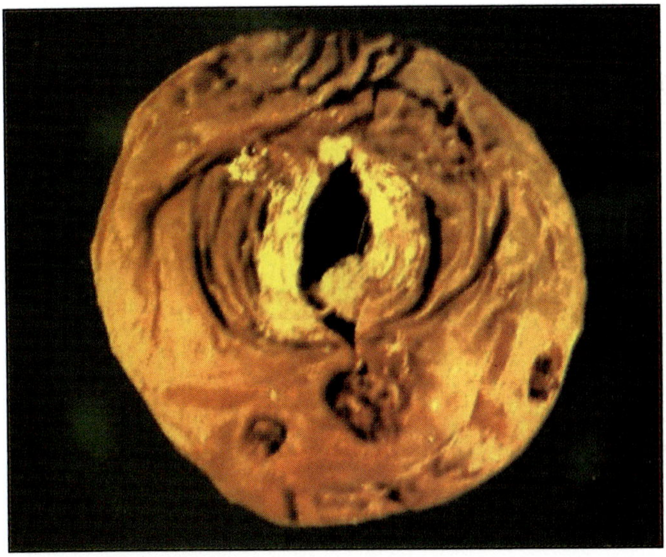

Pathogen

Fusarium oxysporum Schlechtend (Fr.), *F. solani* (Mart) sacc. and other *Fusarium* spp.

Management

1. Avoid bruises to potato tubers by careful handling at harvesting, grading and transport since cuts and bruises predispose the tubers to dry rot infections.
2. Avoid planting of cut tubers if this cannot be avoided treat the cut tubers with dithiocarbamate fungicides before planting to prevent fusarial wilt developing in the succeeding crop.

Wart of Potato

R.K. Arora

Central Potato Research Station, Jalandhar, Punjab

Symptoms

Pathogen

Synchytrium endobioticum (Schilberszky)

Management

1. Cultivate wart immune varieties such as Kufri Jyoti, Kufri Bahar, and Kufri Sherpa.
2. Enforce strict quarantine legislation in affected areas.
3. Practice sanitation and long crop rotations with non hosts such as maize, beans and raddish

Powdery Scab of Potato

R.K. Arora

Central Potato Research Station, Jalandhar, Punjab

Symptoms

Pathogen

Spongospora subterranean (Wallr.) Lagerheim

Management

1. Use disease free seed
2. Avoid flooding of fields
3. Follow crop rotations with non-solanaceous crops

Charcoal Rot of Potato

R.K. Arora

Central Potato Research Station, Jalandhar, Punjab

Symptoms

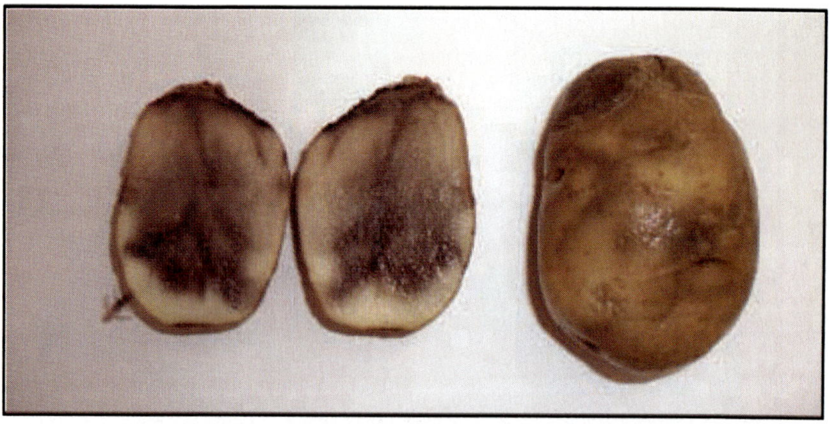

Pathogen

Macrophomina phaseolina (Tassi) Goidanich

Management

1. Plant early maturing cultivars and harvest before soil temperature exceeds 28°C
2. Use disease free seed
3. Use frequent irrigations to keep down the soil temperature
4. Avoid cuts and bruises at harvest

Bacterial Wilt of Potato

R.K. Arora

Central Potato Research Station, Jalandhar, Punjab

Symptoms

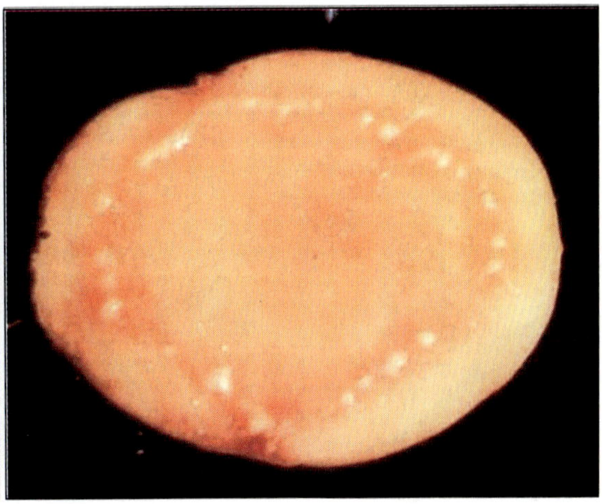

Pathogen

Ralstonia solanacearum (Smith) Yabunchi *et al.*

Management

1. Follow an integrated approach involving use of disease free seed from disease free areas
2. Practice soil solarization in infested area to check soil borne inoculum
3. Follow crop rotation with non hosts such as cereal crops
4. Apply bleaching powder @ 12kg/ha mixed with fertilizers or as soil drenching after first earthing up

Bacterial Soft Rot of Potato

R.K. Arora

Central Potato Research Station, Jalandhar, Punjab

Symptoms

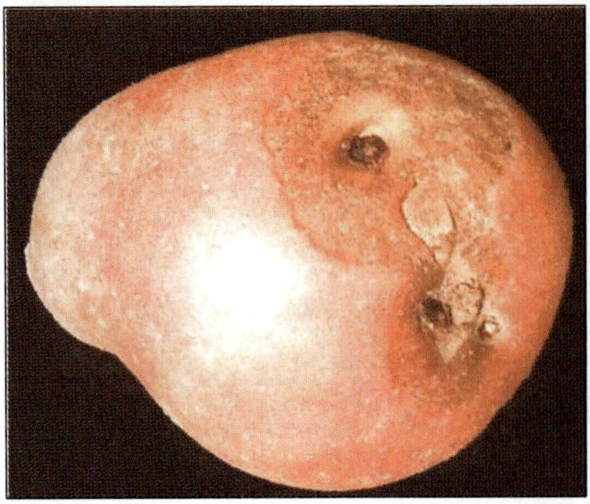

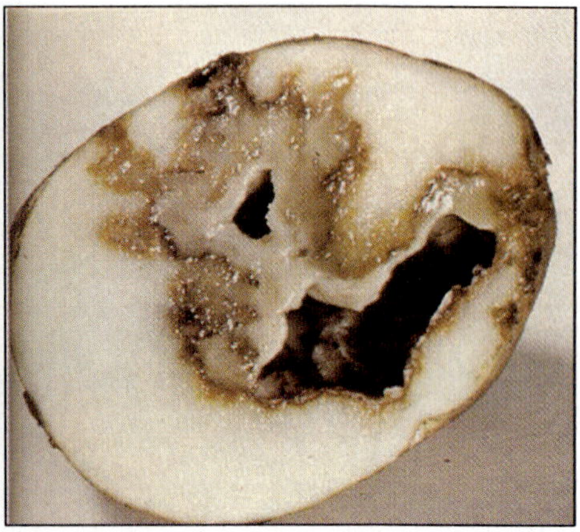

Pathogen

Erwinia carotovora ssp. *caratovora* (Jones) Bergey

Management

1. Plant crop in well drained soil and harvest when the soil is not too wet
2. Avoid bruises and cuts at harvest and subsequent handling.
3. Provide proper aeration during storage and transport.
4. Follow crop rotation with wheat and green manure

Common Scab of Potato

R.K. Arora

Central Potato Research Station, Jalandhar, Punjab

Symptoms

Pathogen

Streptomyces scabies (Thaxter) Lambert and Loria, and other spp. of *Streptomyces*

Management

1. Use acidic fertilizers such as ammonium sulphate
2. Practice soil solarization
3. Treat seed with 3 per cent boric acid
4. Provide frequent irrigation from tuber initiation stage to crop maturity
5. Follow crop rotation with maize, cotton, wheat, cabbage or onion

Potato Wilt

Srikant Kulkarni, Venkatesh Kulkarni and Mamatha M. Gorawar

University of Agricultural Sciences, Dharwad -580 005, Karnataka

Symptoms

Dark brown lesions appear on the stem just below the soil surface. This is followed by drooping and wilting of infected leaves and gradually the whole plant is wilted. Such wilted plants show white cottony fungal thread like and girdling of basal part of the stem. Whitish mycelial growths along with sclerotial bodies, which resemble mustard seeds, are observed at the collar region. Sclerotia are white in the beginning later turn to brown colour. White fungal growth is seen on tubers, which later start rotting and are covered by sclerotia of fungus.

Infection on Collar Region

Infected Field

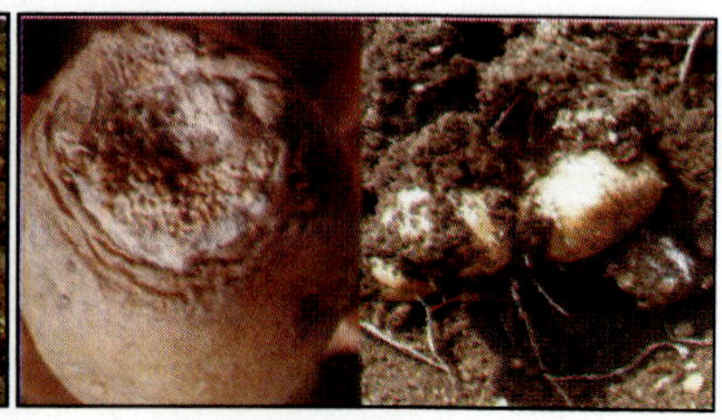

Infected Tubers

Pathogen

Sclerotium rolfsii

Management

1. Seed tuber treatment with *Trichoderma harzianum* @ 10 g/Kg tubers.
2. Soil application of 1 kg *Trichoderma harzianum* with 1 ton of FYM is recommended.
3. Removal and destruction of infected plants.
4. Soil drenching with (0.2 per cent) Carboxin around the infected plants.
5. Crop rotation with rabi sorghum for minimum of two years.
6. Storing of tubers in well-ventilated rooms.

Rice

False Smut of Rice

Yashoda R. Hegde and Srikant Kulkarni

Department of Plant Pathology, University of Agricultural Sciences, Dharwad – 580 005 Karnataka

Symptoms

Symptoms can be observed only after flowering. In the beginning, infected grains turn into a gall like velvety mass known as "Psuedomorph" which become greatly enlarged at later stages. In the beginning they are yellow and covered by a membrane which burst as a result of further growth and color of the ball become orange and later yellowish green and greenish black. Number of smut balls per panicle varies from one to twenty and size from 5 mm to 15 mm in diameter. When smut ball is cut open, it is white in the center and consists of tightly woven mycelium with the glumes and other tissues of the host. There are two outer layers each at a different stage of development. The inner layer is yellowish with radiating mycelium and spores. Outer layer is green and consists of mature spores together with remaining fragments of mycelium.

Figure 1: Yellow Smut

Pathogen

Teleomorph: *Claviceps oryzae sativae* Hashioka (Figure 1)

Anamorph: *Ustilaginoidea virens* (Cke.) Takahashi (Figure 2)

Management

1. Use of healthy seeds.
2. Seed treatment with Carbendazim @ 0.2 per cent.

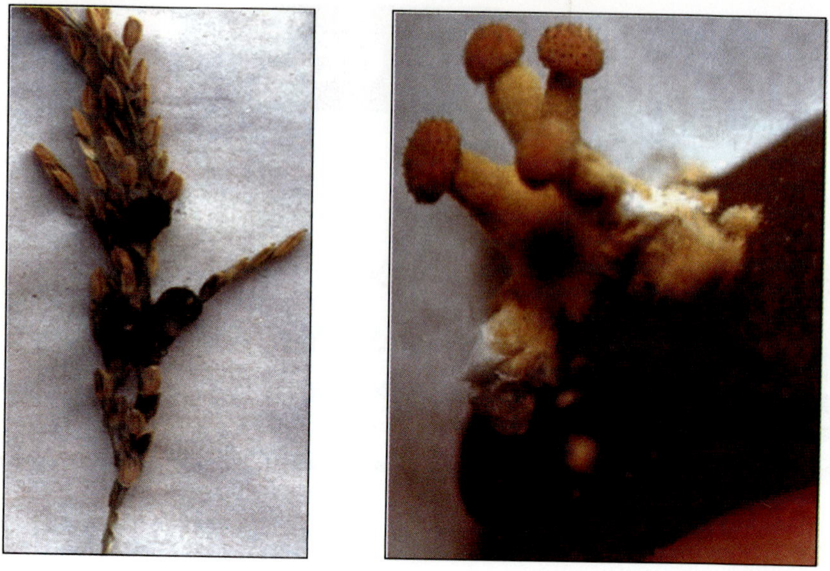

Figure 2: Black Smut Germination of Sclerotia

3. Removal and destruction of infected panicles.
4. Spraying of carbendazim at 0.1 per cent at panicle emergence stage.
5. Grow resistant varieties.

Charcoal Rot of Soybean

G.K. Gupta

NRC for Soybean, Indore – 542017

Symptoms

Seed develops black discolouration. Infected seedlings exhibit reddish brown discolouration at the point of emergence of the hypocotyls. Discolouration in stem is evident at the soil level and above. In well grown plants, typical development of charcoal rot show gray or silvery discolouration of epidermal and subepidermal tissues and light brown discolouration of internal tissues of lower stem and upper tap root. Leaves become chlorotic and wilting is apparent. Production of abundant minute black sclerotia beneath the outer cuticle is a diagnostic symptom of the disease.

Figure 1: Charcoal Rot Infected Plants

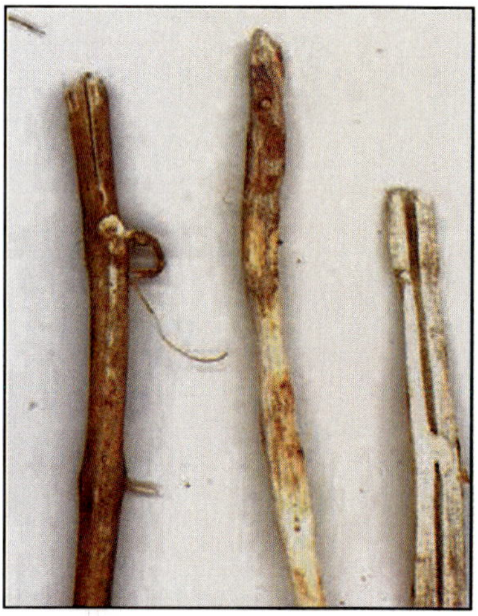

Figure 2: Grey or Silvery Discolouration of Lower Stem and Light Brown Discoloration of Internal Tissues

Pathogen

Macrophomina phaseolina.

Management

1. Use of Resistant Varieties

Cultivation of moderately resistant varieties like NRC 2, NRC 37, JS 71-05, LSb 1 etc.

2. Alteration of Crop Practices

Deep summer ploughing, crop rotation or mixed cropping with cotton or cereals help in minimizing the infection from soil borne inoculum.

3. Biological Control

Use of *Trichoderma harzianum* or *T. viride* for seed treatment @ 10 g/kg seed.

4. Chemical Control

Seed treatment with thiram + carbendazim 1:2 @ 3 g/kg seed or with a combi-product of thiram and carboxin (Vitavax Power) @ 2 g/kg seed.

5. Integrated Management

Timely use of recommended fertilizer (20 N, 60-80 P, 20 K and 20 S kg/ha), maintenance of low plant population to have vigorous plants, flooding of field 3-4 week before sowing or maintaining high soil moisture by irrigation, if possible, during crop growth along with above mentioned management practices.

Collar Rot or Sclerotial Blight of Soybean

G.K. Gupta

NRC for Soybean, Indore – 542017

Symptoms

Light brown-to-brown lesions formed at collar region or just below the soil surface on account of fungus attack results in wilting of seedlings. Most characteristic symptom is the formation of white cottony mats of mycelium around the collar region with reddish brown to dark brown sclerotia about the size of mustard seeds. In older plants too pathogen causes decay in collar region with white mat of mycelium and sclerotia resulting in drooping and wilting of older plants.

Figure 1: Figure White Cottony Mycelium of Fungus at Collar Region

Figure 2: Sclerotia on Stem

Pathogen

Sclerotium rolfsii

Management

1. Use of Resistant Varieties

Cultivation of less susceptible variety NRC 37.

2. Alteration of Crop Practices

Pathogen perpetuate through soil, crop residues and weeds therefore, deep summer ploughing to bury debris and sclerotia to a depth of 15 to 25 cm help in reducing the disease. Crop rotation with maize, sorghum etc. helps to contain the disease to some extent.

3. Biological Control

Seed treatment with *Trichoderma viride* @ 10 g/Kg seed.

4. Chemical Control

Seed treatment with thiram + carbendazim (2:1) @ 3 g/Kg or with a combi product of thiram and carboxin @ 2 g/kg seed. In disease hot spot areas soil drench with Chloroneb @ 20 kg/ha or neem cake @ 2.5 t/ha.

5. Integrated Management

Along with above management practices, sanitation of field, sowing at optimal depth and early sowing escape the infection and roguing and burning of infected plants to check the spread of disease.

Anthracnose and Pod Blight of Soybean

G.K. Gupta

NRC for Soybean, Indore – 542017

Symptoms

Infected seed becomes shriveled, mouldy and brown. Pre-and post emergence damping off/mortality may occur when infected seeds are planted. Otherwise, crop is attacked at all stages of growth but symptoms are evident in the early reproductive stage on stems, petioles and pods. Irregular reddish to dark brown areas appear on infected parts first on the stem. Later on these are covered by black fungal fruiting

Figure 1: Post-emergence Seedling Mortality

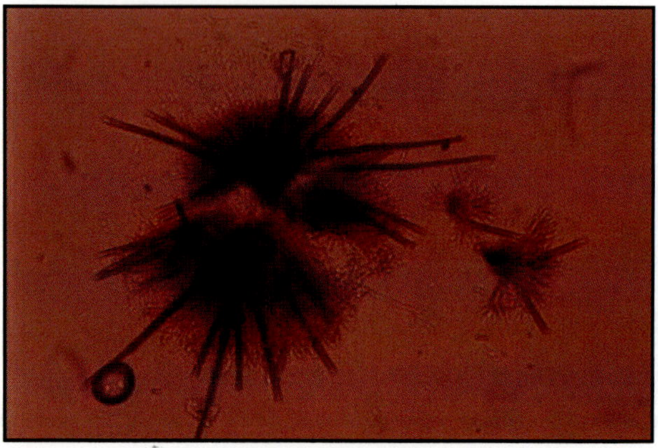

Figure 2: Dark Brown Areas with Acervuli on Stem and Pods

bodies (acervuli) with setae (minute black spines), which can be seen by unaided eye. These setae are diagnostic character of the disease. Foliar symptoms are expressed in the form of laminar veinal necrosis, leaf rolling and defoliation but under prolonged period of high humidity. Affected plants also senesce earlier. Pods turns yellow to brown, seed formation is also affected as they become shriveled and mouldy, and sometimes seeds do not form in the pods

Pathogen

Colletotrichum dematium f.sp. *truncatum* or *Colletotrichum truncatum*.

Management

1. Use of Resistant Varieties

Cultivation of moderately resistant varieties like Bragg, Himso 1563, Hardee, PK 472, JS 80-21, Pusa 37, VLS 21, NRC 12 etc.

2. Alteration of Crop Practices

Early sowing to escape the disease.

3. Chemical Control

Seed treatment with thiram +carbendazim (2:1) or captane @ 3g/Kg seed and spray of carbendazim or thiophanate methyl @ 0.05 per cent mancozeb 0.2 per cent on infected crop

4. Integrated Management

Use of clean and healthy seeds, burning of infected plant debris in combination with other above mentioned disease management practices.

Rust of Soybean

G.K. Gupta

NRC for Soybean, Indore – 542017

Symptoms

Initially abundant chlorotic gray brown minute spots of the size of tip of sewing needle appear in-group on the leaves, primarily on the lower surface. Slowly these spots increase in size and turn to tan to reddish brown and angular measuring 2 to 5 mm^2 and later form pustules. Usually the leaf tissues around the group of spots become yellow which distinguishes rust symptoms from bacterial pustule symptoms where leaf tissues around a single spot become chlorotic and yellow. Leaves turn brown within a short time causing early defoliation and reduction in number of pods, seeds and seed weight. Presence of loose brown powder owing to rupture of pustules is a characteristic symptom.

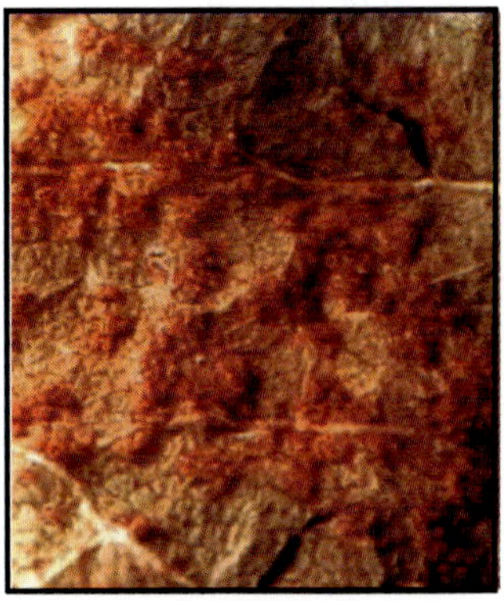

Figure 1: Rust Pustules

Pathogen

Phakopsora pachyrhizi

Figure 2: Chlorotic and Gray Brown Minute Spots of Rust

Figure 3: Rust infected Brown Prematurely Dried Leaf

Management

1. Use of Resistant Varieties

Cultivation of rust tolerant varieties like Ankur, PK 1024, PK 1029, JS 80-21, Indira soybean 9, MAUS 61-2 or early maturing varieties.

2. Alteration of Crop Practices

No summer and rabi cultivation of soybean and rouging of self sown soybean plants in these seasons. Intercropping of soybean with other rainy season crops of the area.

3. Chemical Control

Two to three sprays of hexaconazole (Contaf) or propiconazole (Tilt) or triadimefon (Bayleton) or oxycarboxin (plantvax) @ 0.1 per cent are found effective. To enable entering soybean field for spray at least after every 15 rows a strip of about 1.50 m be left vacant in the field.

4. Integrated Management

Deep summer ploughing, roguing and burning of infected plants, crop and crop residues during the season and one protective spray of any above fungicides over the crop at 35 to 40 DAS in rust hot spot areas should additionally be tried for proper management of the disease.

Cercospora Blight or Leaf Spot and Purple Seed Stain of Soybean

G.K. Gupta

NRC for Soybean, Indore - 542017

Symptoms

Infected seeds may have pink to light purple to dark purple areas ranging from small specks to large blotches covering whole surface. Upon germination shriveled and discoloured cotyledons are produced which fall prematurely and resulted seedlings may die with in a few days. Usually disease appear at or after the full pod stage in the form of reddish-brown to purple, angular to irregular lesions ranging from tiny spots to large irregular blotches on leaves. They may coalesce to form large necrotic areas. Blighting of younger upper leaves over large areas is the striking symptom of the disease. Veinal necrosis, defoliation from upper young leaves along with early senescence of the plant is also very common. Reddish purple to reddish black slightly sunken lesions are also formed on stem, petiole and pod and seeds. Heavily infected stems have a dull gray to dark brown appearance and dry up 7 to 10 days prematurely.

Figure 1: Purple Stained Infected Seeds

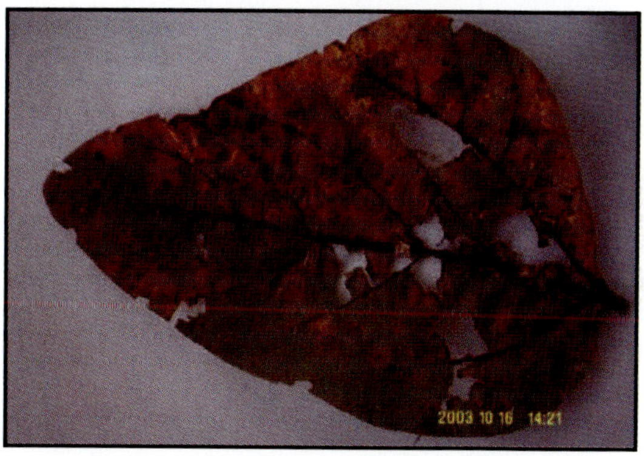

Figure 2: Reddish-brown to Purple, Angular to Irregular Lesions of the Pathogen

Pathogen

Cercospora kikuchii.

Management

1. Use of Resistant Varieties

Cultivation of less susceptible varieties like JS 80-21 and Bragg etc in hot spot areas.

2. Chemical Control

Seed treatment with thiram + carbendazim (2:1) or captane @ 0.3 per cent and subsequent spray of benomyl @ 0.1 per cent or thiophenate methyl@ 0.05 per cent over infected crop.

3. Integrated Management

Use of clean and certified seeds along with above-mentioned disease management practices.

Myrothecium Leaf Spot of Soybean

G.K. Gupta

NRC for Soybean, Indore – 542017

Symptoms

Infected seeds when sown cause seed rot and blight of emerged seedlings. As such, disease is visible around 30 to 35 days after planting as small round brown spots on leaves. Later on these spots become dark with dark brown or purple margin surrounded by a translucent area in concentric rings. Slowly spots merge with each other and become of irregular shape. On maturity, sporodochia are visible as white erumpent structures on translucent zone of the spots, which later turns to black dot like structures on leaves as well as on pods. This is the characteristic symptom of the disease.

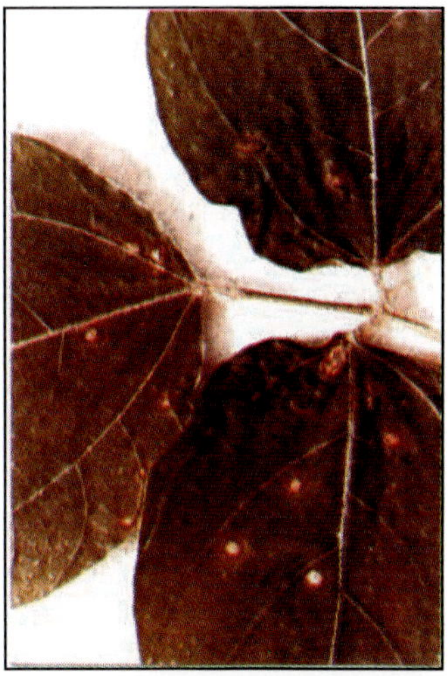

Figure 1: Spots with Dark Brown to Purple Margin Surrounded by a Translucent Area on Leaves

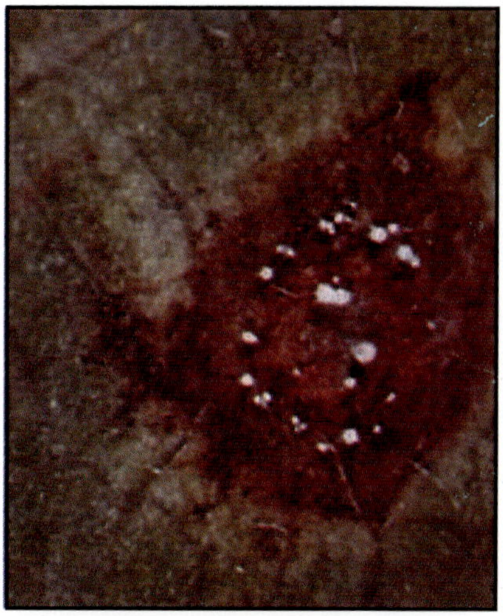

Figure 2: White Erumpent Sporodochia on Translucent Zone of the Spots

Pathogen

Myrothecium roridum

Management

1. Use of Resistant Varieties

Cultivation of resistant/moderately resistant varieties like Bragg, JS 71-05, JS 335, MACS 13, MACS 124, MAUS 47, NRC 7, PK 564, etc.

2. Chemical Control

Seed treatment with thiram + carbendazim (2:1) @ 3 g per kg followed by spray of carbendazim or thiophanate methyl (0.05 per cent to 0.1 per cent or 400 to 800 g/ha) or mancozeb (0.25 per cent or 2 kg/ha). First spray at 35 DAS is very effective.

3. Integrated Management

Use of healthy and certified seed along with above management practices.

Root Rot, Stem Rot and Foliage/ Aerial Blight of Soybean

G.K. Gupta

NRC for Soybean, Indore – 542017

Symptoms

Seedlings from infected seeds may have infected shoot apices resulting in stunted growth. On seedlings, reddish brown sunken lesions often surrounded by brown mycelial growth are formed on roots and basal portion of the stems, which later causes post-emergence mortality. Water soaked spots that subsequently turn reddish brown to dark brown appear first on lower leaves. Later on, entire leaf gets blighted in due course. Root rot of adult plant in the form of brown to dark brown discolouration of the cortical and lignified tissues is also common. Development of a reddish-brown canker around the base of the stem, easy peeling off cortical tissues of the root and drooping of leaves which remain attached to the dead plant are the other characteristics of this disease. Under high rainfall or high humid conditions a web

Figure 1: Reddish Brown Sunken Lesions on Root and Lower Stem

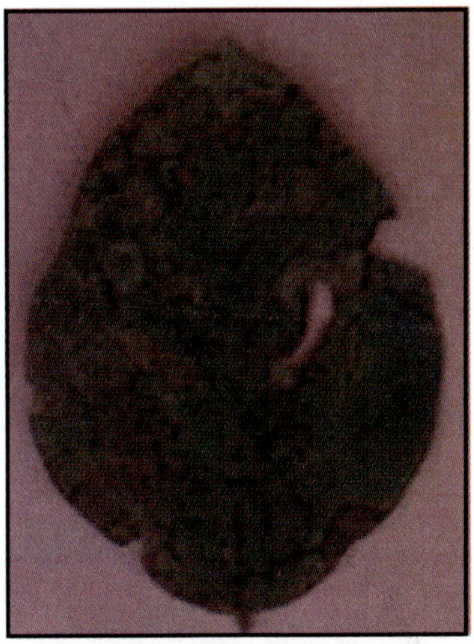

Figure 2: Water Soaked to Reddish Brown Spots on Leaf

like mycelium growth of the fungus forms on the leaves and in between the closely spaced infected plants. Dark brown sclerotia are formed on leaves and petioles.

Pathogen

Rhizoctonia solani.

Management

1. Use of Resistant Varieties

Cultivation of moderately disease resistant/tolerant varieties like PK 262, PK 416, PK 472, PK 1042, PK 564, SL 295 etc., specially for aerial blight.

2. Alteration of Crop Practices

Avoidance of excess plant population and monoculture.

3. Chemical Control

Seed treatment with thiram + carbendazim (2:1) @ 3 g or combi product of thiram + carboxin @ 2 g or captan @ 3 – 4 g/kg. One spray of carbendazim or thiophanate methyl (0.05 per cent) or mancozeb (0.25 per cent) is found very effective.

Bacterial Pustule of Soybean

G.K. Gupta

NRC for Soybean, Indore – 542017

Symptoms

Disease appears as minute, pale green elevated spots on interveinal areas on both sides of the leaves. Later on small raised light gray to yellowish pustules are formed in the center of spots usually on the undersurface with a pale green halo around them. The pustules may coalesce in to large spots. Leaves become ragged when dead areas are torn away by wind. Spots may also develop on pods of susceptible variety.

Figure 1: Minute, Pale Green Elevated Spots on Leaf

Figure 2: Leaf with Raised Light Gray to Yellowish Pustules in the Center of Spots

Pathogen

Xanthomonas axonopodis pv. *glycines* Syn. *Xanthomonas campestris* pv. *glycines.*

Management

1. Use of Resistant Varieties

Cultivation of resistant varieties like PK 327, PK 416, PK 1029, PK 1042, JS 71-05, JS 90-41, Himso 1563, Indira soya 9, KHSb 2, MAUS 32, NRC 7, NRC 37, SL 96, VLS 2 etc.

2. Chemical Control

Seed treatment with Streptocycline @ 500 mg/kg seed and spray of copper-oxychloride (1.6 kg) + Streptocycline (160 g) or Kasugamycin/Validamycin (1.6 kg) + copper-oxy chloride (1.6 kg) in 800 litre of water/ha.

3. Integrated Management

Use of disease free and clean seed harvested from healthy plants for cultivation, destruction of diseased crop residues/debris and deep summer ploughing along with above-mentioned practices manage the disease.

Soybean Mosaic

G.K. Gupta

NRC for Soybean, Indore – 542017

Symptoms

Infected seeds many times get mottled by having the same colour on seed coat as that of hilum, popularly known as hilum bleeding. When infected seeds are used for sowing, the initial symptoms of the disease are severe mosaic of dark and light green areas and downward curling of primary leaves giving seedlings spindly appearance. Such plants have rugosed, crinkled, mottled and curled unifoliate leaves. Subsequently trifoliate leaves become chlorotic, severely stunted, mottled and rugosed. Trifoliate leaves of well-grown plants have a mosaic of light and dark green areas that may

Figure 1: Black Hilum Bleeding on Infected Seeds

Figure 2: Raised or Blistered Light and Dark Green Areas on Trifoliate

later become raised or blistered mostly along the main veins. Dark green enations along the puckered leaves and necrosis are also occasionally present Necrosis, leaf yellowing, terminal necrosis and defoliation at times lead to death of Plant. Diseased pods often remain stunted, flattened and curved with less pubescence and fewer and small sized seeds.

Pathogen

Soybean mosaic virus causes the disease. The virus is a flexuous rod belongs to potyvirus group and is seed, graft and sap transmitted. Disease has a narrow host range and being transmitted in a non-persistent manner by about 32 aphid species belonging to 15 different genera. The disease is occurring all over the soybean growing areas of India. It can reduce yields ranging from 50 to 93 per cent. The extent of damage to crop depends on age of plant at the time of infection. Symptoms are more severe with cool temperature (near 18°C) and humid condition but less severe at 24-25°C, while symptoms are masked above 30°C.

Management

1. Use of Resistant Varieties

Use of moderately resistant varieties like JS 71-05, KHSb 2, LSb 1, MACS 58, MACS 124, PK 416, PK 564, Punjab 1, VLS 2 etc.

2. Alteration of Crop Practices

Cultivation of 2 to 3 or more varieties and change of varieties at least after every two years.

3. Chemical Control

Pre sowing soil application of phorate @ 10kg/ha., Two foliar sprays of thiamethoxam 25 WG @ 100 g/or methyl demeton 800ml/ha at 30 and 45 days after sowing.

4. Integrated Management

Weed free clean cultivation in addition to above management practices keep the crop relatively free from the virus.

Yellow Mosaic of Soybean

G.K. Gupta

NRC for Soybean, Indore – 542017

Symptoms

Symptoms can be seen in seedlings with in 15 days of germination as small yellow diffused patches that later on coalesce to make leaf more or less complete yellow. Normally, symptoms of the disease appear in the form of conspicuous yellowing along small veins of the leaves followed by severe yellow mosaic and mottling of leaves, which is a diagnostic symptom. Yellowing is either scattered or produced in indefinite bands along the major veins. Rusty necrotic spots appear in the yellow areas as the leaves mature.

Yellow Mosaic Infected Plant

Pathogen

Mung bean yellow mosaic virus: It is a gemini virus recently named genus *Begomovirus*.

Management

1. Use of Resistant Varieties

Use of resistant varieties like PK 416, PK 472, PS 564, PK 1024, PK 1029, PK 1042, Pusa 37, SL 295, SL 525 etc.

2. Alteration of Crop Practices

Cultivation of 2 to 3 or more varieties and change of varieties at least after every two years.

3. Chemical Control

Pre sowing soil application of phorate @ 10kg/ha. Seed treatment with thiamethoxam 70 WS @ 3g/kg seed in hot spot areas and spray of thiamethoxam 25 WG @ 100 g or methyl dematon 25EC @ 0.8 lit/ha or ethofenprox 10 EC @ 1.0 lit/ha.

4. Integrated Management

Weed free cultivation and rouging of infected plants and plant parts in addition to above-mentioned practices are promising for the management of disease.

Indian Bud Blight of Soybean

G.K. Gupta

NRC for Soybean, Indore – 542017

Symptoms

Figure 1: Browning and Curving of the Terminal Bud to Form a Crook

Most striking symptom is the browning and curving of the terminal bud to form a crook. Later on other buds become brown, necrotic and brittle. Necrosis of veins, midrib, petiole and stem is also common. Petioles are often thickened and shortened and may be curved distorting the shoot tips. Sometimes the petioles become so brittle that they break by pressing with fingers. Leaflets are dwarfed and tend to cup or roll and the leaf blades become rugosed and bronzed. If infection is early then plants become severely stunted with shortened internodal length and several auxillary shoots. Pods are not formed or remain severely underdeveloped or aborted.

Pathogen

In India it is *Peanut Bud necrosis virus* whereas in other countries it is *Tobacco ring spot virus.*

Figure 2: Crop with Brown, Necrotic and Brittle Auxillary Buds

Management

1. Use of Resistant Varieties

Less susceptible varieties *viz.* JS 71-05, JS 79-81, KHSb 1 etc. should be cultivated.

2. Chemical Control

Seed treatment with Imidacloprid @ 5 g/kg and spray of nuvacron (0,01 per cent) or rogor (0.01 per cent) or any other systemic insecticides.

3. Integrated Management

In addition to above mentioned measures crop should be kept free of broad-leaved weeds and rouging and burning of disease infected plants or plant parts should be practiced.

Sugarcane

Whip Smut of Sugarcane

V.C. Khilare

Department of Botany, Vasantrao Naik Mahavidyalaya, Aurangabad – 431 003, M.S.

Symptoms

1. The earliest symptom is the grass-like appearance of shoots, which have small and narrow leaves. Later, long, black whip-like structures appear at the stalk tip. Longer whips will curve downwards.
2. Plants are stunted and the shoot is converted into black whip like structure.
3. The shoot consists of black powder of smut spores.
4. The black shoot is covered with white thin cover.
5. After bursting, the whip smut spores are distributed all over the field.
6. After cutting the sugarcane, the spores are present on the stumps.
7. The spores germinate on the stumps of ratoon crop.

Pathogen

Ustilago scitaminea Sydow

Management

1. Use the sugarcane setts from disease free fields.
2. Collection of whips in the bag and burning away from the field.
3. Treatment of setts with hot water (50°C) at least one hour before cultivation.
4. Dip the cane setts in fungicides solution like carbendazim or tridimefon @ 1 per cent eliminates fungal propagules in the seed cane. Such treatments eliminate smut and other fungal colonization in the seed canes.
5. Use resistant varieties like CoH 119, Co 98014, Co 99006, Co 8371, Co 85004, Co 86032, Co 87025, Co 87044, Co 91010, CoM 88121, Co 94008.

Red Rot of Sugarcane

V.C. Khilare

Department of Botany, Vasantrao Naik Mahavidyalaya, Aurangabad – 431 003, M.S.

Symptoms

1. On leaves, there is a reddish streak on the midrib.
2. Internal tissues of the stalk have a red coloration, which is interrupted by elongated white patches.
3. Diagnostic symptoms can best be observed by splitting the stalk lengthwise. The infected tissues have a dull red colour interspersed by occasional white patches across the stalk.
4. These white patches are specific to the disease and are of significance in distinguishing red rot from other stalk rots.
5. During monsoon, the disease spreads fast and the whole crop dries and not a single millable cane is obtained.

Pathogen

Colletotrichum falcatum

Management

1. Soaking of sugarcane setts in a 25 per cent solution of fungicide like carbendazim/thiophanate methyl for 24 hrs before planting was found to be effective.

2. Biocontrol strain strains of *Pseudomonas fluorescens* were effective against red rot pathogen.
3. The incidence of red rot can be reduced through good cultural practices, such as clearing fields of excessive trash and efficient drainage.
4. Avoidance of planting susceptible cultivars during cool and wet weather is effective. Regular rouging of diseased plants, burning of trash, ploughing affected fields, maintenance of proper soil moisture, and prompt harvesting of infected or susceptible crops are other management practices recommended. Bunding of affected field should be done to avoid movement of rain or floodwater.
5. Ratooning of diseased crop should be avoided. Diseased crop should be harvested as early as possible. Crop rotation should be followed in affected fields.
6. Resistant varieties are Co 86010, Co 86011, Co 86249, Co 93009, Co 94008 Co 8021, Co 86249, Co 93009, BO 91, Co 89003, Co 96015, Co 97009, Co Pant 94211.

Sunflower

Downy Mildew of Sunflower

Srikant Kulkarni and Yashoda R. Hegde

Department of Plant Pathology, University of Agricultural Sciences, Dharwad – 580 005, Karnataka

Symptoms

Symptoms can be observed on all the parts at different stages of plant growth. In seedling stage, this disease can result in damping off of seedlings wherever there is heavy inoculum and favorable conditions for disease. Seedlings can be killed either before emergence or immediately after emergence. The systemic infection result in downy growth of fungus consisting of sporangiophore and sporangia, on lower surface of leaves. On the corresponding upper surface of the leaves appear chlorotic yellow patches. Severely infected plants failed to produce flower heads. Even if flower heads are produced, they seldom produce seeds. Systemically infected plants exhibited typical erect head, with loss of phototropic responses normally seen in healthy sunflower plant. Major influence of systemic infection was extreme stunting depending on the stage at which infection occurred. The root infection may result in

Figure 1: Downy Growth

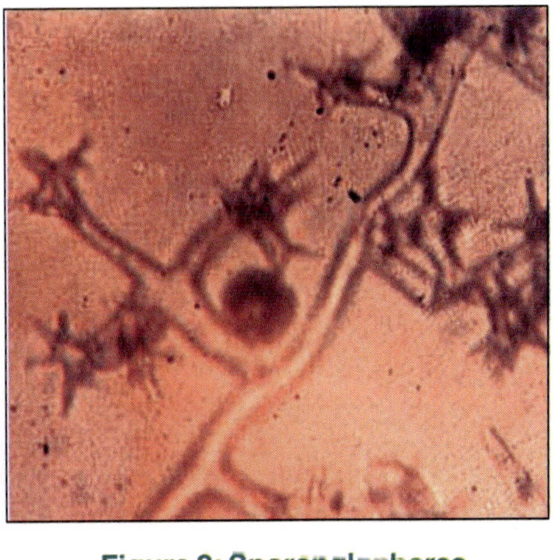

Figure 2: Sporangiophores

formation of root galls at the base of primary roots. Such roots are scurfy, hypertrophied and with reduced fibrous roots.

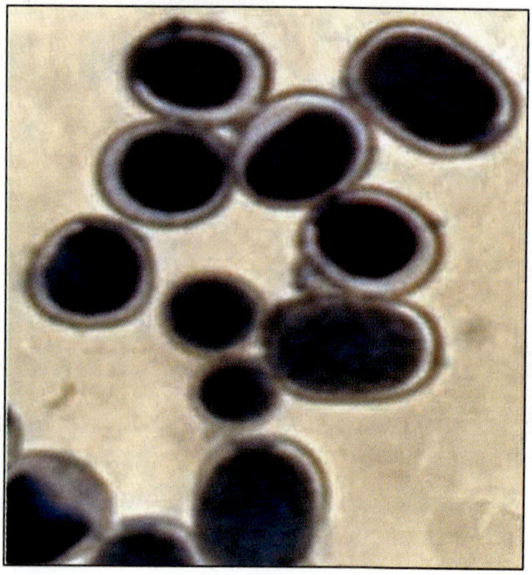

Figure 3: Sporangia

Pathogen

Plasmopara halstedii (Farl.) Berl. and De Toni.

Management

1. Use of healthy seeds. Avoid the use of seeds from diseased area.
2. In India, race of pathogen identified is European race-1. The other virulent races known outside the country should be prevented by means of strict quarantine measures.
3. Deep sowing (>3 cm) increases disease incidence and delays emergence, which increases the opportunity of infection to the pathogen. Hence, shallow sowing is desirable.
4. Late planting particularly in July- August should be avoided.
5. Eradication of infected plants.
6. Chemical: Seed treatment with metalaxyl (Apron 35 SD) @ 6 g/kg seed.
7. Growing of resistant varieties *viz.*, LDMRSH-1, LDMRSH-3, MLSFH-93, DSH-4, SH-491.

Tomato

Early Blight of Tomato

Y.L. Nene

Asian Agri-History Foundation, Secunderabad – 500 009

Symptoms

1. Symptoms can be seen on leaves, stems, and fruits.
2. Older leaves first show small, brown to dark brown lesions. Yellow discoloration can be seen around the lesions.

3. When lesions on leaves increase in number, complete leaves may turn yellow and ultimately fall off.
4. If brown lesions are examined closely, concentric rings can be seen clearly in the large lesions.
5. Lesions on stems are small, dark brown, and sunken. When lesions increase in size, these become circular or elongated. Concentric rings can be seen in the enlarged lesions.
6. Green or ripe fruits can be affected. Lesions on fruits can be large and usually show concentric rings. With a hand lens, black powdery mass of fungal spores can be seen. Fruit drop is common.

Pathogen

Alternaria solani (Ell. and Mart.) L.R. Jones and Grout

Management

The disease is favored by warm, wet weather. High soil fertility reduces disease severity. The fungus also attacks potato {*Solanum tuberosum* L.) and eggplant (*Solanum melongena* L.; brinjal). The fungus survives in the diseased crop debris and on seed. Control can be obtained through the use of healthy seed, long crop rotations, resistant cutivars, and fungicides.

Fusarium Crown and Root Rot of Tomato

Y.L. Nene

Asian Agri-History Foundation, Secunderabad – 500 009

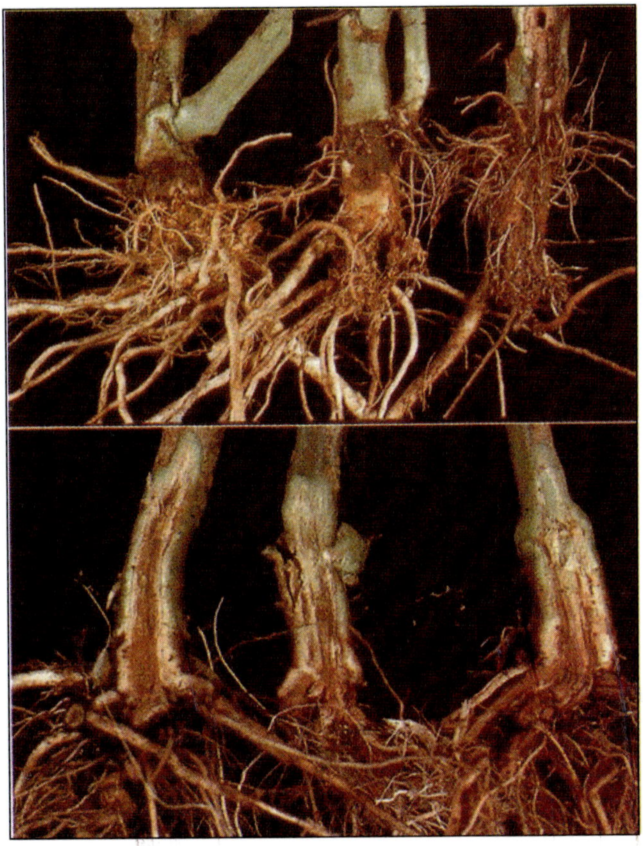

Symptoms

1. Symptoms can be seen on roots, stems, and leaves.
2. Symptoms on leaves usually show up after fruiting has started. The oldest leaf first shows yellowing on margins. The yellowing is followed by browning and the petiole collapses. Younger leaves are then affected likewise.
3. Whole plant may show wilting.
4. Affected stem shows brown, necrotic lesion at and above soil surface. Stem cankers may be seen.
5. Root system is affected. Roots and stems show brown discoloration internally.

Pathogen

Fusarium oxysporum Schlechtend.:Fr. f. sp. *radicis-lycopersici* W.R. Jarvis and Shoemaker.

Management

The disorder has been observed more commonly in the greenhouse crop than in the field crop. Control can be obtained through steaming of greenhouse soil and then drenching with a fungicide. Also sterilizing the soil with a fumigant is effective.

Late Blight of Tomato

Y.L. Nene

Asian Agri-History Foundation, Secunderabad – 500 009

Symptoms

1. The disease can occur in an epidemic form as it can spread rapidly.
2. All above-ground parts of plants are affected.
3. On leaves, symptoms appear as water-soaked spots, which enlarge rapidly into pale green to brown lesions. Under moist conditions, whitish fungal growth can be observed on lesions on the underside of leaves. Death of leaves occurs rapidly.
4. Petioles and stems are affected with lesions and may die in a short time.
5. Symptoms on fruit are expressed in the form of dark, greasy spots, which can enlarge to occupy the entire fruit. Complete rotting of fruits can be observed.
6. Foul odor can be experienced in a badly affected crop.

Pathogen

Phytophthora infestans (Mont.) de Bary

Management

Wetness, cool nights (12-15 C), and moderately warm days (20-25°C) favor the disease. Computer software to forecast disease is available. This fungus can attack potato and it survives in the diseased crop debris. Control can be obtained through destruction of infected crop debris and through application of fungicides.

Sclerotium Stem Rot of Tomato

Y.L. Nene

Asian Agri-History Foundation, Secunderabad – 500 009

Symptoms

1. Plant is prone to attack in all stages under favorable environment.
2. Stem is most commonly attacked and the main symptom is brown to black rotting of stem at the soil level.
3. Seedlings fall on the ground due to weakened stem and adult plants show wilting.
4. If the soil is wet, white mycelium of the fungus can be seen on the diseased portions.
5. Later, mustard-seed-like sclerotia of the fungus can be seen on the mycelial mats.
6. Fruits in contact with the soil can be attacked. Water-soaked, soft, star-shaped lesions appear and the affected fruit collapses in 3-4 days. The lesion cavity shows white mycelium with sclerotia.

Pathogen

Sclerotium rolfsii Sacc.

Management

The fungus can attack a large number of plant species, but some cereal crops such as maize (*Zea mays* L.) and millets are not affected. Wet soil and high temperatures (above 30°C) favor the disease. The fungus survives in soil for many years as sclerotia and keeps itself alive as a saprophyte. The disease can be managed through soil solarization, biological control with *Trichoderma harzianum* Rifai, and crop rotations with non-susceptible crops.

Bacterial Spot of Tomato

Y.L. Nene

Asian Agri-History Foundation, Secunderabad – 500 009

Symptoms

1. All above-ground plant parts are affected by the bacterium.
2. Lesions, generally brown and circular are seen on leaves, stems, and fruits.
3. Lesions of this disease on leaves can be confused with those of early blight or gray leaf spot. Bacterial spot lesions do not have concentric rings (such rings exist in case of early blight) and are not distributed uniformly over the leaf surface (gray leaf spot lesions are uniformly distributed).
4. Lesions may coalesce and blighting of foliage can occur.
5. Usually there is no halo around lesions on fruits.

Pathogen

Xanthomonas campestris pv. *vesicatoria* (Doidge) Dye

Management

Frequent rain or sprinkler irrigation with warm range of temperatures (25-30°C) favor the disease. The bacterium is seedborne, soilborne, and also survives on other plant species. Control can be obtained through seed treatment, crop rotation, and application of bactericides.

Bacterial Wilt of Tomato

Y.L. Nene

Asian Agri-History Foundation, Secunderabad – 500 009

Symptoms

1. Initial wilt symptoms appear in a few young leaves as these become flaccid due to lack of moisture.
2. A rapid and complete wilting follows.
3. If the disease progress (*i.e.*, wilting) is slow, adventitious roots appear on the stems.
4. If an infected stem is cut horizontally, droplets of yellowish, sticky ooze will be visible on the cut surface.
5. The bacterial wilt can be easily distinguished from the fungal wilts. Just take a freshly cut piece of stem and let the cut end remain in water in a dish. Milky exudate oozing out from the cut end will confirm bacterial wilt.
6. Wilted plants eventually show rotting of roots.

Pathogen

Pseudomonas solanacearum (Smith) Smith

Management

The bacterium attacks a large number of plant species. Injuries in roots allow the bacterium to enter plants. The bacterium survives in soil. Control can be obtained through the use of resistant cultivars.

Tomato Mosaic

Y.L. Nene

Asian Agri-History Foundation, Secunderabad – 500 009

Symptoms

1. Leaves show mottling.
2. Early infection leads to stunted, chlorotic plants.
3. Leaves may show upward curling or downward rolling, depending upon the strain of virus that infects the plant.
4. Sometimes, high temperature (+30°C) may result in masking (disappearance) of symptoms.

Pathogen

Tomato mosaic virus

Management

Tomato mosaic virus has several strains. The disease spreads from one plant to another by workers' hands, implements, and clothing, since the virus is highly infectious. The virus is seedborne. Control can be obtained through the use of resistant cultivars and use of seed treatment with chemicals such as trisodium phosphate.

Root-knot of Tomato

Y.L. Nene

Asian Agri-History Foundation, Secunderabad – 500 009

Symptoms

1. Large populations of nematodes in the root system leads to stunting of plants leading finally to premature wilting.
2. Secondary symptoms of nutrient deficiencies may be observed.
3. The most characteristic symptom is the appearance of swellings, called galls or root-knots, all over the root system. These are usually brownish but may become dark.

Pathogen

Meloidogyne incognita (Kofoid and White) Chitwood

Management

The nematode is active throughout the year in warm, moist soils. The nematode can attack a large number of plant species. Control can be obtained through crop rotation with cereal crops, soil solarization or sterilization, and use of resistant cultivars.

Bacterial Wilt of Brinjal and Tomato

P.A. Fugro and K.P. Vaidya

Department of Plant Pathology, Dr. B.S. K. K. V., Dapoli – 415 712, Dist. Ratnagiri, M.S.

Symptoms

The first symptom of the disease in the field is dwarfing of the affected plants and bronze discolouration of the leaflets. The characteristic symptom of the disease is sudden wilting of the plants. On cutting across the infected root of the plant, whitish bacterial ooze can be seen under low power of microscope. Vascular bundles are found to be discoloured dark brown. The disease can attack the plant at any stage of the crop growth except in nursery.

Pathogen

Ralstonia solanacearum

Management

Spray antibiotics like streptomycin sulphate, streptomycin, etc. Some biocontrol agents like *Pseudomonas fluorescence* have also been proved to be encouraging but not with uniform and guaranteed success. However, the manipulation of cultural practices and use of different agro techniques have not yielded quite successful results. Indian Horticultural Research Institutes, ICAR, State University Research Stations and private firms have evolved some of the high yielding bacterial wilt resistant varieties. Some of these are:

Brinjal: Arka Nilkuntha, Arka Keshav, Arka Nidhi, Swarna Prabha, BB60-e, etc.

Tomato: Sonali, Arka Alok, BWR-1, SUN-7611, SUN-7610, etc.

Alternaria Blight of Tomato

P.A. Fugro and K.P. Vaidya

***Department of Plant Pathology, Dr. B.S. K. K. V.,
Dapoli – 415 712, Dist. Ratnagiri, M.S.***

Symptoms

Disease appears generally 20-30 days after transplanting in the field. *Alternaria* blight generally produces leaf spots as the most noticeable signs of the disease during early stages of blight development. The spots enlarge and coalesce covering larger area of the leaf. The disease spreads on entire aerial parts of the plant. Brownish elongated legions produced on the stem can girdle the base of the stem killing or stunting the younger plants and many times inducing the wilting of older plants. The flowers and fruits also show the sunken concentrated brownish sports, which enlarge and cause premature fall of the affected lot. The losses may be as high as 80 per cent depending upon the stage of the crop during which infection occurs.

Pathogen

Alternaria solani (Ell. and Martin) Jones and Grout.

Management

Seed should not be collected from the plots where disease has been occurred. Diseased plant debris should be collected and destroyed by burning. Under konkan conditions raising nursery beds during the month of October and transplanting the

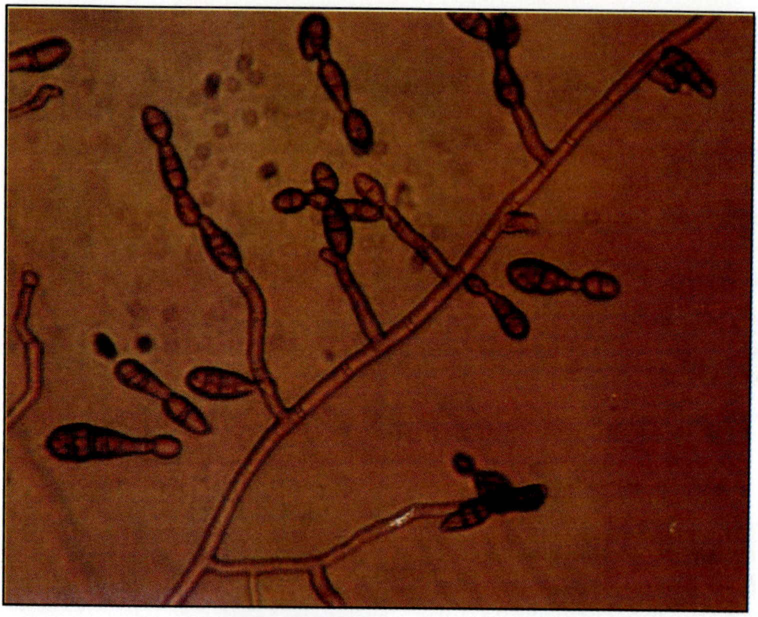

crop before 15th November (47th meteorological week) has given satisfactory yield. However, repeated spraying of the crop (3-4 times) with Mancozeb (0.25 per cent) or Chlorothalonil (0.25 per cent) at an interval of 10-12 days one month after transplanting has been found to be useful in minimizing the disease incidence and maximization of yield. Most of the high yielding varieties including the varieties recommended for this region have been found to be susceptible to early light. However, some of the varieties coming from reputed private firms as SUN-7611, SUN-7610 and varieties evolved by Dr. B.S.Konkan Krishi Vidyapeeth, Dapoli as DPL-T-14, DPL-T-9, have shown remarkable field tolerence to *Alternaria* blight. Besides, these varieties are also resistant to bacterial wilt.

Wheat

Flag Smut of Wheat

A.K. Sharma, J. Kumar and M.S. Saharan

Directorate of Wheat Research, Aggarsain Marg, Post Box 158, Karnal, Haryana – 132001

Symptoms

Pathogen

Urocystis agropyri (Preuss) Schroet. (syn. *U. tritici* Koern.)

Management

1. Seed treatment: Carboxin (75 WP) @ 2.5 g/Kg seed
2. Agronomic practices: Early and shallow sowing reduces incidence

Powdery Mildew of Wheat

A.K. Sharma, J. Kumar and M.S. Saharan

Directorate of Wheat Research, Aggarsain Marg, Post Box 158, Karnal, Haryana – 132001

Symptoms

Powdery Mildew on Leaf

Powdery Mildew on Spike

Pathogen

Erysiphe graminis DC. f.sp. *tritici* E. Marchal

Conidial chain of *Erysiphe graminis*

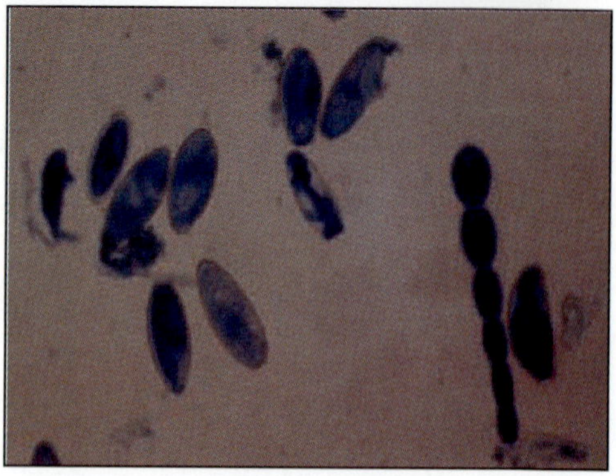

Conidial Chain

Management

Chemical spray: Karathane 0.05 per cent for 3 - 4 times after appearance of symptoms or one spray of propiconazole (25 EC@ 0.1 per cent).

Foliar Diseases of Wheat: Spot Blotch, Leaf Spots and Blights

A.K. Sharma, J. Kumar and M.S. Saharan

Directorate of Wheat Research, Aggarsain Marg, Post Box 158, Karnal, Haryana – 132001

Symptoms

Figure 1: Leaf Blight

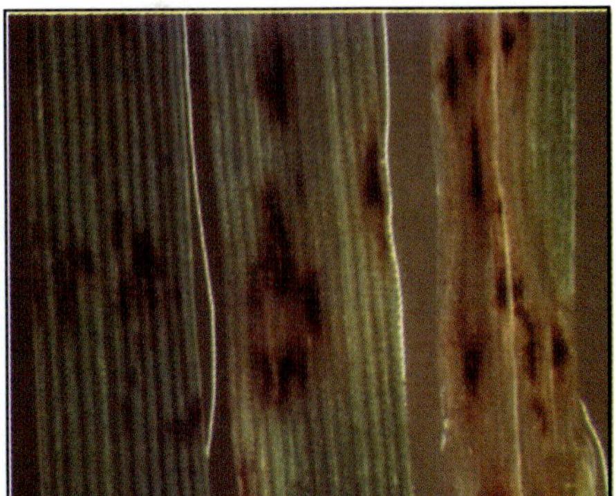

Figure 2: Spot Blotch

Pathogen

Bipolaris sorokiniana

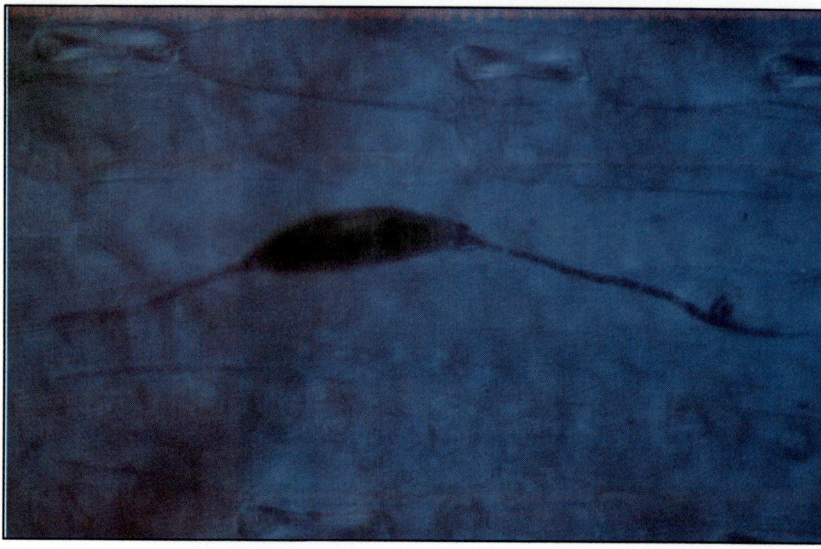

Figure 3: Germinating Conidium of *B.sorokiniana* on Leaf Surface

Management

Foliar spray of propiconazole (25 EC @ 0.1 per cent) at disease appearance and at 15 - 20 days interval if need be. CIMMYT stocks Suzhoe, Chirya and Myna, Yang Mai 6 as well as *Triticum aegilopoides*, *T. monococcum* and *T. timopheevi* are promising breeding stocks for resistance.

Head Scab/Head Blight of Wheat

M.S. Saharan, J. Kumar and A.K. Sharma

Directorate of Wheat Research, Aggarsain Marg, Post Box 158, Karnal, Haryana – 132001

Symptoms

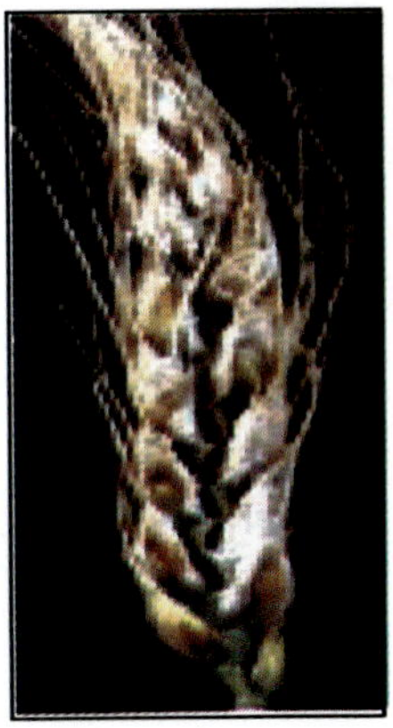

Pathogen

Fusarium graminearum Schwabe

Fusarium graminearum **Macroconidia**

Management

No effective chemical control method available since the disease is of limited importance in India. Sumai 3 is a promising stock for resistance breeding.

Hill Bunt of Wheat

A.K. Sharma, J. Kumar and M.S. Saharan

Directorate of Wheat Research, Aggarsain Marg, Post Box 158, Karnal, Haryana – 132001

Symptoms

(*Source*: Compendium of Wheat Diseases, APS Press, The American Phytopathological Society)

Pathogen

Tilletia foetida (Wallr.) Liro and *T. caries* (DC) Tul.

Management

Seed treatment with Thiram @ 2-3 gm/kg seed.

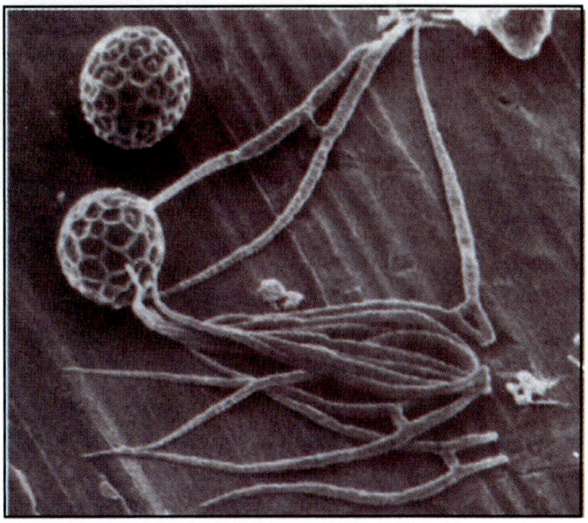

Germinating Teliospores of *Tilletia* sp. (*Source*: Compendium of Wheat Diseases, APS Press, The American Phytopathological Society)

Karnal Bunt of Wheat

J. Kumar, A.K. Sharma and M.S. Saharan

Directorate of Wheat Research, Aggarsain Marg, Post Box 158, Karnal, Haryana – 132001

Symptoms

Figure 1: Spike with Diseased Grains

Figure 2: Diseased Grains

Pathogen

Tilletia indica Mitra, Synonym: *Neovossia indica* (Mitra) Mundkur (Figures 3–5).

Management

1. Resistant Varieties

Aestivum: PBW 154, WL 1562, HD 2281,PBW 502 Durum: PBW 34, DWL 5023, PDW 215, PDW 233.

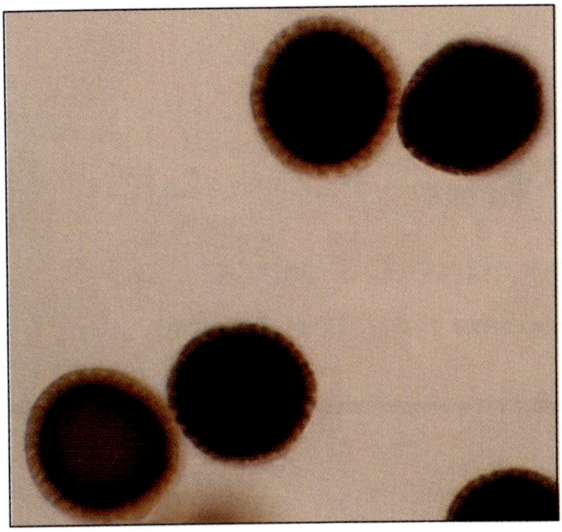

Figure 3: Teliospores

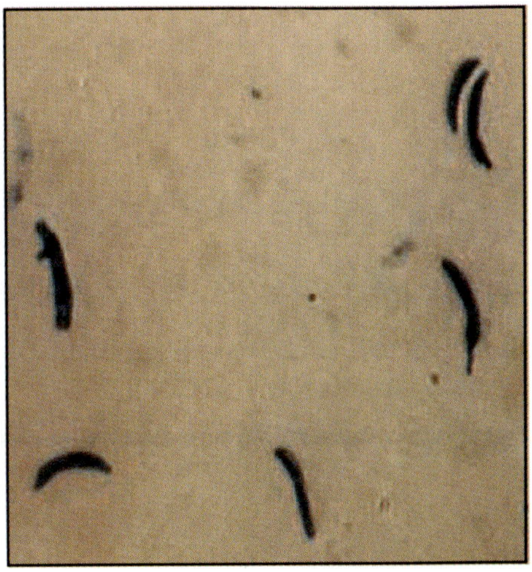

Figure 4: Allantoid Sporidia

2. Chemical Control

Propiconazole 25EC @ 500ml/ha gives an effective control.

3. Agronomic Approaches

Follow timely sowing (15th November), Avoid irrigation at Flowering and sprinkler irrigation in particular, Biological mulch of chickpea and Polythene mulch after tillering, Non cultivation of wheat for two consecutive years, Soil solarization for 21 – 28 days during summers using plastic sheets.

Figure 5: Germinating Teliospore

4. Biocontrol

One spray of propiconazole (25EC) @ 0.1 per cent may be given (in seed crop only) at ear head emergence stage. Two sprays of *Trichoderma viride*, one at Zadoks' growth stage 31 – 40 and another at 41 – 49 gives non chemical disease management at par with one spray of propiconazole (25EC). One spray of *T. viride* at crop growth stage 31 – 39, followed by one spray of propiconazole (25 EC) @ 0.1 per cent at growth stage 41 – 49 gives cent percent disease control.

Molya Disease of Wheat

A.K. Singh, A.K. Sharma and J. Kumar

Directorate of Wheat Research, Aggarsain Marg, Post Box 158, Karnal, Haryana – 132001

Symptoms

Figure 1: Susceptible (S) and Resistant (R) Plants

Pathogen

Heterdera avenae Woll (Syn *H. major* (Schmadit) Franx) (Figure 2).

Management

1. Crop rotation with mustard, pulses or fenugreek or carrot for one or two years reduces the nematode population.

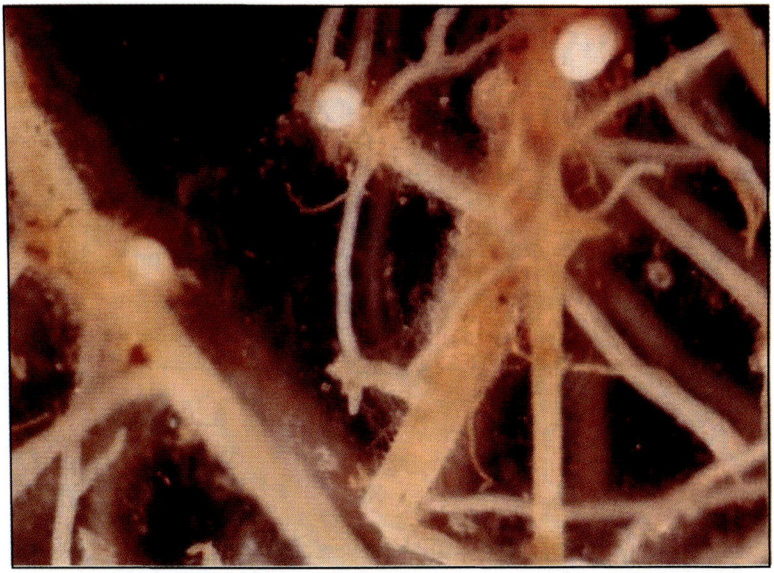

Figure 2: Pearly Root Caused by *Heterdera avenae*

2. Introduction of wheat and barley resistant varieties like Raj MR-1, RD 2035 and RD 2052
3. Summer ploughing in May-June can control the nematode.
4. Soil application of Carbofuran @ 1.5 kg a.i./ha at before sowing.

Ear Cockle Disease of Wheat

A.K. Singh, A.K. Sharma and J. Kumar

Directorate of Wheat Research, Aggarsain Marg, Post Box 158, Karnal, Haryana – 132001

Symptoms

Figure 1: Infected Ear Head and Grains

Pathogen

Anguina tritici (Steinb) Chit.(Figure 2).

Management

1. Sieving of seeds to remove seed galls.

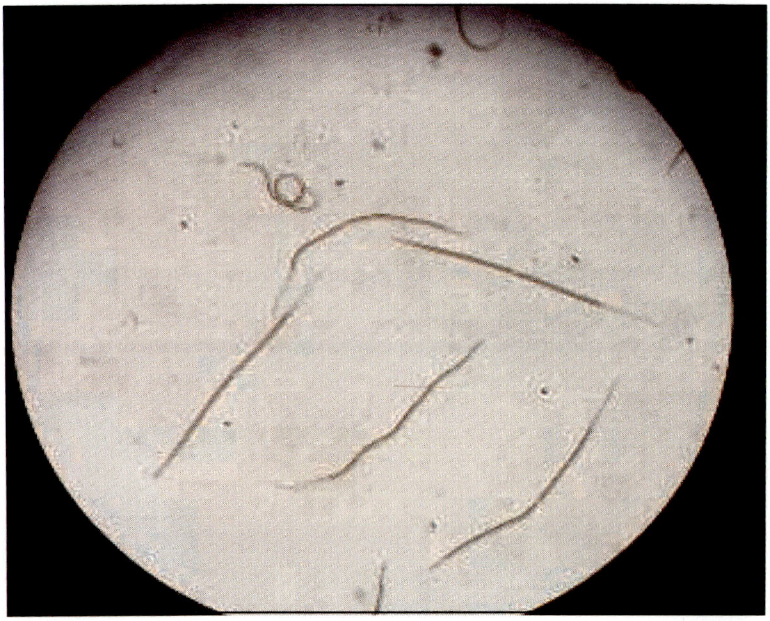

Figure 2: Larvae of *Anguina tritici*

2. Brine solution (2-5 per cent of common salt) floatation technique to remove galls from the seed lot.

Loose Smut of Wheat

A.K. Sharma, J. Kumar and M.S. Saharan

Directorate of Wheat Research, Aggarsain Marg, Post Box 158, Karnal, Haryana – 132001

Symptoms

Figure 1: Loose Smut of Wheat

Pathogen

Ustilago segetum (Bull) *var. tritici* (Figure 2).

Management

Seed treatment with carboxin (75WP@ 2.5 g/Kg seed) or carbendazim (50WP@ 2.5 g/Kg seed) or tebuconazole (2DS @ 1.25 g/Kg seed) or a combination of a reduced

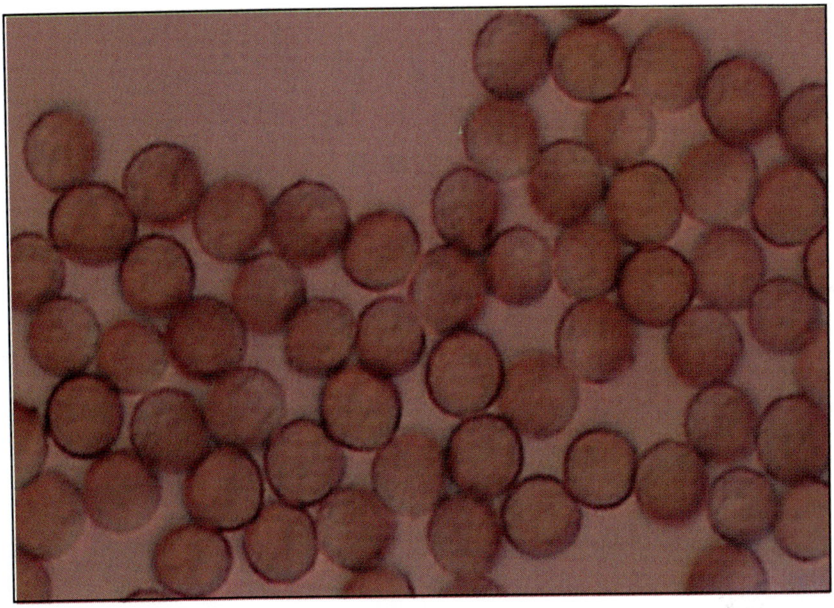

Figure 2: Chlamydospores

dosage of carboxin (75WP@ 1.25 g/Kg seed) and a bioagent fungus *Trichoderma viride* @ 4 g/Kg seed.

Wheat Rusts: Stem (Black), Leaf (Brown), Stripe (Yellow)

A.K. Sharma, J. Kumar and M.S. Saharan

Directorate of Wheat Research, Aggarsain Marg, Post Box 158, Karnal, Haryana – 132001

Symptoms

Figure 1: Brown Rust

Figure 2: Black Rust

Figure 3: Yellow Rust

Pathogen

Stem (black) rust of wheat: *Puccinia graminis* Pers. f.sp. *tritici* Eriks. and Henn.

Leaf (brown) rust of wheat: *P. recondita* Rob. Ex. Desm. f.sp.*tritici* Eriks. and Henn.

Yellow rust of wheat: *P. striiformis* Westend.

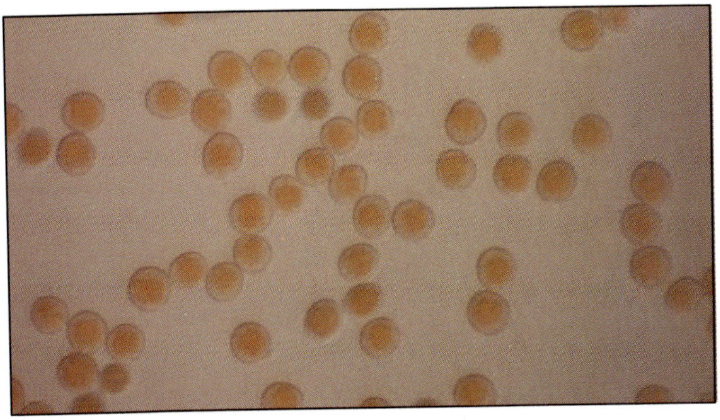

Figure 4: *P. recondita tritici* Uredospores

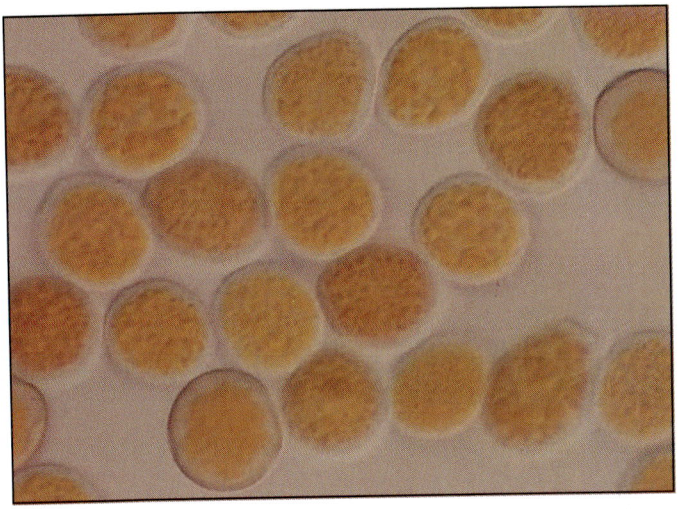

Figure 5: *P. striiformis* Uredospores

Management

1. The most economical method is to plant resistant cultivars as follows:
2. North West plain zone: DBW 17, PBW 343, PBW 502, RAJ 3765, WH 542, WH 896, UP 2425, PBW 373, WH 896, PDW 233 (durum) etc.
3. North east plain zone: PBW 343, HD 2733, K 9107, NW 1014, NW 1012, HUW 533,K 8804, DL 784-3, HP 1633 etc.
4. Central zone: HI 1500, HI 8627, HW 2004, HD 4672, HI 8638, DL 788-2, GW 273, GW 173, HI 8498 (durum) etc.
5. Peninsular zone: GW 1189(d),NIDW 295(d), GW 322, HD 2189, HUW 510, Lok 45, NIAW 34, MACS 1967(d), MACS 2846(d), RAJ 4037(d), MACS 2496, DWR 162, DWR 106 (durum), DDK 1001 and DDK 1009 (dicoccum) etc.
6. North hill zone: HPW 42, HS 420, SKW 196, VL 616, VL 804, VL 829, VL 830 etc.
7. Southern hill zone: HW 971

Index